About the Editors

Robert Stobaugh is Charles E. Wilson Professor of Business Administration at Harvard Business School. He spent eighteen years in industry in the United States and abroad before earning his doctorate at Harvard Business School. Author, coauthor, or coeditor of nine books in three subject areas—international business, technology and production, and energy—he is a past president of the Academy of International Business. Stobaugh's best-known book, of which he was a principal author and coeditor, is *Energy Future: Report of the Energy Project at the Harvard Business School.* His most recent research has been on the innovation of new products and processes and the subsequent spread of technology.

Louis T. Wells, Jr., is Herbert F. Johnson Professor of International Management at Harvard Business School. He has been doing research on multinational firms based in developing countries and on government structures and procedures to negotiate with foreign investors. His recent publications include *Third World Multinationals* and an article on "Bargaining Power of Multinationals and Host Governments."

D1279961

OCLC

Technology Crossing Borders

*The Choice,
Transfer, and
Management
of International
Technology
Flows*

The Choice, Transfer, and Management

HBS

PRESS

Contributors

Michel Amsalem
Henri de Bodinat
James Keddie
Donald J. Lecraw
Farshad Rafii
Robert Ronstadt
Robert Stobaugh
Piero Telesio
Louis T. Wells, Jr.
David Williams
Wayne Yeoman

Technology Crossing Borders

of International Technology Flows

564948

Edited by
Robert Stobaugh
and
Louis T. Wells, Jr.

Foreword by Michael Y. Yoshino

Harvard Business School Press

Boston, Massachusetts

Faculty research at the Harvard Business School is undertaken with the expectation of publication. In such publication the Faculty member responsible for the research project is also responsible for statements of fact, opinions, and conclusions expressed. Neither the Harvard Business School, its Faculty as a whole, nor the President and Fellows of Harvard College reach conclusions or make recommendations as results of Faculty research.

Harvard Business School Press, Boston 02163
© 1984 by the President and Fellows of Harvard College.
All rights reserved.
Printed by Halliday Lithograph in the United States of America.
88 87 86 85   5 4 3 2

Chapter 3 is a revised version of Louis T. Wells, Jr., "Economic Man and Engineering Man: Choice of Technology in a Low Wage Country," *Public Policy*, Vol. XXI, No. 3, Summer 1973, John Wiley & Sons. © 1973 by John Wiley & Sons. Used with permission.

Chapter 6 was adapted from Michel Amsalem, *Technology Choice in Developing Countries*, Cambridge: MIT Press, 1983. © 1983 by the Massachusetts Institute of Technology. Used with permission.

Chapter 9 is based on material from Piero Telesio, *Technology Licensing and Multinational Enterprises*, New York: Praeger Publishers, 1979. © 1979 by Praeger Publishers. Used with permission.

Chapter 11 was adapted from Robert C. Ronstadt, "International R&D: The Establishment and Evolution of Research and Development Abroad by Seven U.S. Multinationals," *Journal of International Business Studies*, Spring–Summer 1978. © 1978 by the Academy of International Business. Used with permission.

Library of Congress Cataloging in Publication Data

Main entry under title:

Technology crossing borders.

   Includes bibliographic references and index.
   1. Technology transfer—Addresses, essays, lectures.
I. Stobaugh, Robert B. II. Wells, Louis T.
T174.3.T375 1984   338.9   83-26591
ISBN 0-87584-158-9

To Ray, who was much help;

To Josie, who, regardless of her position, was too.

Contents

Part 2: Channels for Transferring Technology

Part 3: Management of Technology

Foreword

This book is dedicated to Raymond Vernon, our mentor, colleague, and friend. Raymond Vernon has received worldwide acclaim for his scholarly contributions in several areas. In this volume, we honor him for his achievements in the field of multinational enterprise.

During the almost two decades of his association with the Harvard Business School, Vernon organized and directed two major research projects on multinational enterprises, one dealing with U.S.-based firms, the other with their European and Japanese counterparts. It was his intellectual leadership that brought all of us together, literally from the four corners of the earth. To the study of multinational enterprises, he brought his intellect, his characteristic zeal, and his training in economics, tempered by years of experience in government and business.

Out of these two projects came a number of major studies that shed considerable light on the behavior of the multinational enterprise. Vernon's own books, *Sovereignty at Bay* and *Storm Over the Multinationals,* are classics in the field. Recognizing the futility of trying to encompass all of his contributions in the field, we have chosen international transfer of technology as the theme of this book. It is a topic that has been a central focus of his research and an area in which many of his students have made important contributions.

All of us who are associated with this volume had the privilege of studying under a truly remarkable man, a man of awe-inspiring intellect combined with an uncompromising commitment to the pursuit of excellence. On many occasions he has chided us for loose thinking, pushed us for sharper analysis, and challenged us to greater creativity. He is a tough taskmaster indeed; but at the same time, we know of no one who is so generous of his time and ideas. Beneath his intellectual toughness is a genuine, warm, and concerned person, and these are the qualities we have come to appreciate most in him.

It is only fitting that we should dedicate this volume to Raymond Vernon on the occasion of his retirement from the Business School, with our appreciation, admiration, and above all affection.

Michael Y. Yoshino

Professor of Business Administration
Harvard Business School

Acknowledgments

In the course of researching and writing this book, we and our colleagues pestered hundreds of business managers, government officials, labor leaders, academics, and other specialists. Some of them requested anonymity, and in any case space limitations would make it impossible to name them all individually. But we express our deepest appreciation to these people, who made our work possible.

Fortunately there are others whom we can thank by name. The research was carried out during the tenure of two deans at the Harvard Business School, Lawrence Fouraker and John H. McArthur, whose support was essential to us and to our colleagues. Also we thank the Division of Research, first under Richard Walton, next under Richard Rosenbloom, and then under E. Raymond Corey and Joanne Segal. In addition to the support provided by the Division of Research, financial help for individual pieces of research came from the Ford Foundation and the Harvard Institute for International Development. Besides the editors of this volume, William Abernathy, Steven Allen, Tom Allen, Richard Caves, Mel Horowitch, Maurice Kilbridge, Frederick Knickerbocker, Richard Rosenbloom, Bruce Scott, Wickham Skinner, Yoshi Tsurumi, and, of course, Raymond Vernon were involved in supervising work done as doctoral theses.

Several secretaries helped type various manuscripts. In addition, Cathy Judson and Beverly Davies were invaluable in the overall coordination of the project.

Max Hall provided us with detailed editorial assistance on some of the studies and overall editorial advice on the project, as he has on so many other occasions. Nancy Jackson and Mary Barno edited all the chapters. We also thank David Otte, Judy Uhl, and Molly Mulhern for launching this book, the first to be published by the new Harvard Business School Press.

With so much help, one might think we would have a perfect book. In fact we don't, but that's our own fault.

Robert Stobaugh
Louis T. Wells, Jr.

1

Introduction

**ROBERT STOBAUGH AND
LOUIS T. WELLS, JR.**

Technology imported from other nations has become a major source of know-how for many countries; international sales of technology represent significant income for multinational enterprises and their home countries. Decisions about such flows affect firms' profits and nations' balances of payments, economic growth, employment levels, and even security and independence. Nevertheless, relatively little has been published about how technology transfers are managed.

The empirical work in this book examines three issues in the transfer of technology:

1. How managers, public and private, choose the kinds of technology they import or export;

I

2. How multinational enterprises decide on the channels through which they transfer technology and how that choice affects the recipient firm abroad; and
3. How multinational enterprises manage certain of their relationships with overseas affiliates that import, use, modify, and generate technology.

This book makes no attempt to summarize all the literature in the fields on which it reports; rather, it presents a group of clearly related empirical studies that draw on a common set of concepts. On some points the studies are in agreement with the conventional literature; on others, they depart strikingly from the more commonly accepted theories.[1]

The themes in the studies derive, to a great extent, from the researchers' common views of the behavior of firms. Most of the writers assume that management attention, like capital, is a scarce resource to be allocated to activities that yield the greatest returns. Further, they all reject the assumption that enterprises operate in perfectly competitive markets in favor of the assumption that technology is chosen, transferred, and managed in industries that are more or less oligopolistic. Firms differ as to how they choose to compete in these industries, and in different industries and countries they face different institutional arrangements—labor laws, financing possibilities, and so on. The structure of the market, a firm's position in the market, and the external environment affect the choices and decisions that are the subjects of this book.

All the studies draw at least part of their analytic approach from conventional economic theory. Yet, because their main purpose is to analyze certain activities in the real world, all the authors introduce approaches that are not usually incorporated in strictly economic models. To some extent, the researchers have set up simple economic models as straw persons to be challenged with an arsenal of complex empirical findings. This is especially the case with the studies of choice of technology, which have relied the most heavily on economic models. The models in the early studies seem overly simple to the reader familiar with today's work within the discipline of economics. Yeoman's pioneering research, for example, poses a model of technology choice that nearly all economists would now reject as too simple, though it was the dominant one at the time his work was in progress. To be sure, the literature hinted at changes to come, but these notions had hardly taken hold of the profession,[2] nor had they yet affected the advice offered to policy makers in governments or, for that matter, in firms.

On the other hand, economic models had little to say in the past about some of the other issues examined in these studies. Only recently has the discipline begun to develop models to determine, for example, whether market mechanisms or extensions of the firm are appropriate for the transfer of assets such as technology, or how to manage extensions such as foreign subsidiaries.[3]

CHOICE OF TECHNOLOGY

To the traditional economist, the choice of a manufacturing technique seemed to fit the basic framework of economic theory. Economists generally assumed a world of homogeneous products, many small competitive firms, and managers whose decisions were based on economic objectives. In such a world, firms would be forced to choose the technique that minimized their manufacturing costs. The task of the manager was simply to find the technology that combined the factors of production, usually considered to be capital and labor, in the least costly way.

True, certain engineering questions would have to be resolved before the economic choice could be made: What techniques are feasible, and what are their technical coefficients? But if the engineer provided information about the factors of production required by available manufacturing techniques, and if capital and labor costs were known, the economist could then provide a simple framework that would allow the manager to choose the technology that combined capital, labor, and other factors to yield minimal production costs.

The economist's simple model implied that firms manufacturing the same product in a particular country would all use the same minimum-cost techniques (unless some faced different costs for labor or capital, of course). Any firm that deviated from the optimal choice would incur higher manufacturing costs than its competitors and would consequently be forced by the market to modify its technology or to leave the business. The conventional theory also offered clear predictions of how technology would differ in developing and advanced countries. Faced with low wages and high capital costs, managers in poor countries would choose more labor-intensive techniques than would managers in the advanced countries, and this choice of labor-intensive technology would contribute to solving a potential unemployment problem.

By the early 1960s, it was becoming clear to economists and others that the simple model was inadequate to explain the decisions actually made by managers. Observers noted that firms in the same country were using quite different technologies to manufacture ostensibly identical products. Flouting theory, labor-intensive and capital-intensive plants were surviving side by side in the developing countries; and, contrary to expectations, expanding industrial sectors in developing countries were not creating large numbers of job opportunities. Part of the explanation seemed to be that the technologies used by factories in low-wage countries often were not significantly more labor intensive than those used in the advanced countries, so they did little to relieve unemployment. Again, the predictions of the simple economic theory were confounded by the facts.

Despite the widening gap between the theory and the facts of technology choice, in 1968, when Yeoman completed the first study in this book, "Selection of Production Processes by U.S.-Based Multinational Enterprises" (Chapter 2), there had been virtually no empirical work on how managers actually chose the technology they transferred or how they adapted it. The little work that did exist consisted primarily of some engineering studies, which demonstrated that a rather wide range of techniques was feasible in certain industries, and some research based on aggregate data on factor utilization.

With more sophisticated studies now available, the methodology of Yeoman's study seems rather simple. Yet its contribution and that of Strassmann, which appeared in the same year,[4] are seminal to the ideas developed by other researchers, some of whose work is reported in later chapters. The core of Yeoman's work is captured in his hypothesis that the extent to which U.S. multinationals adapt their technology to foreign costs, especially lower labor and higher capital costs, depends primarily on two variables: the price elasticity of demand faced by the individual firm and the relation of manufacturing costs to total costs. The larger the value of each of these variables, the greater the extent of adaptation.

The tentative confirmation of this hypothesis stimulated others to undertake more rigorous tests of similar hypotheses and to explore why managers' behavior differed so much from the pattern predicted by simple economic theory. The efforts of Wells, Keddie, Lecraw, and Amsalem reported in this book confirm Yeoman's basic contention that the nature and degree of competition faced by the manager is a critical variable in the choice of technology. Moreover, these studies demonstrate that the role of competition in determining manufacturing technique is important regardless of whether the firm is a multinational, a local private firm, or even a state enterprise. Though the range of possible choices of technology varies from one industry to another, most of the factors that influence the decision are similar in all industries.

The second study reported in this book, "Economic Man and Engineering Man" by Wells (Chapter 3), was greatly influenced by Yeoman's (and Strassmann's) work. In contrast to Yeoman, Wells included locally owned firms as well as multinationals. Examining forty-three plants in Indonesia, he confirmed the role of price elasticity of demand in determining manufacturing technology, concluding that severe price competition forced managers to minimize manufacturing costs. This study expanded the explanation of managers' behavior: when they could escape severe price competition, whether through their success in differentiating products or through actions of governments that shielded them from competition, managers did not attempt to minimize costs through their choice of manufacturing technique. Wells proposed two kinds of reasons for the observed behavior:

1. Managers have objectives other than profit maximization, among which are: a) "engineering objectives," which lead them to prefer sophisticated, automated plants over simple, unautomated, labor-intensive factories; and b) "managerial objectives," which include reducing labor problems and producing the highest-quality product, even if consumer response is not sensitive to the quality difference.
2. Managers attempt to reduce the risks of liquidity problems and any errors in matching production capacity to demand by using capital-intensive plants. Payrolls have to be met regularly. On the other hand, bank debt is more flexible in some developing countries; in Indonesia, for example, state banks

would not always press for repayment of loans for equipment if a company fell on hard times. Moreover, contrary to conventional wisdom, the output level of automated plants may be more flexible than that of labor-intensive factories. Labor laws and practices may severely limit layoffs or dismissals. Workers, once slowed, cannot be brought up to speed again, but machines can be cut back and sped up.

Keddie, the author of another study on production techniques in Indonesia (Chapter 4), examined a larger number of firms than did Wells, and he focused more on the issue of product quality than did the other researchers, arguing that more capital-intensive techniques enable a firm to produce high-quality products more reliably than do labor-intensive techniques. For some firms, a strategy of quality and advertising yield a net benefit, even after allowing for the higher costs of capital-intensive manufacturing. In certain industries (such as paint and soap), a combination of high quality and branding provides a way to avoid severe price competition. In Keddie's view, capital-intensive production is an inherent part of the strategy of most firms in such an industry, since the techniques assure high-quality output. In certain other industries, such as printing, weaving, and textile printing, the gains from product differentiation seem insufficient to cover the higher manufacturing costs associated with capital-intensive processes. Here managers' choices lean toward the labor-intensive end of the spectrum in a country such as Indonesia, where labor costs are low. Cigarettes and flashlight batteries fall in an intermediate position, because of the strongly segmented nature of these markets in Indonesia, where branding and price competition survive side by side.

Lecraw, the author of "Choice of Technology in Thailand" (Chapter 5), made a particularly important contribution by collecting data from a large number of firms (400) and by using sophisticated quantitative techniques to analyze these data. Lecraw constructed production functions for Thai industry in order to examine the different techniques employed by various firms in Thailand. The results again supported the role of competition in the choice of technology, but with much more convincing evidence than had been mustered before. Firms that projected high profits before they built plants chose capital-intensive techniques. This was particularly true of firms that followed a strategy of product differentiation and branding and which had high selling costs. Moreover, firms in industries with many competitors leaned toward labor-intensive technology.

Lecraw provided another test of the role of competition by separating exporting firms from those that supplied the domestic market. Firms that exported final products from Thailand, and for which price competition was therefore particularly important, used more labor-intensive techniques than otherwise similar firms that manufactured for the domestic market. Lecraw, like the previous author, emphasized the role of product quality. Quality was important, for example, for firms that exported intermediate products for use by affiliates. Much of the observed labor substitution was in materials-handling operations, where the impact of technology on product quality is minimal.

In addition to his methodological contribution, Lecraw observed that firms managed by engineers chose more labor-intensive techniques than did other firms, a finding that contrasts with what one might infer from the Wells study. If "engineering objectives" are important, it may be that they are most faithfully pursued by nonengineers, a possibility that suggests the need for a change in terminology.

Amsalem, in his study of the textiles and paper industries (Chapter 6), was the first in this group of authors to determine the full slate of technologies from which a firm might select, rather than examining only the technologies in use in the countries in question. He did this by obtaining information on other possible techniques from manufacturers of similar products in the United States and Japan, and from engineers and equipment suppliers in various countries. He was then able to show that one of the barriers to the selection of an optimal technique was managers' ignorance of the full range of techniques available elsewhere.

Amsalem also was the first to pay considerable attention to a process industry, where the range of feasible alternatives was thought to be small. The focus was on pulp and paper (a process industry) and on the spinning and weaving of textiles, in both Asia and Latin America. He studied the very micro level—selected individual steps in the production processes.

Amsalem also explored the influence of several kinds of risk on the manager's choice of technology. Particularly important among these was the risk that human error would affect the quality of output in a process industry, but he also pointed out the risk that a small firm faces when making choices different from those of established industry leaders in oligopolistic industries. Although it may not minimize costs, the firm that follows the leaders at least will have technology as good as that of its rivals, whereas a poor choice of alternative technology can cost a small firm its existence.

Amsalem's is also the first study in this book to make calculations of optimal technology from the point of view of the country as well as that of the firm. He attempted calculations based on the "social optimum" to determine how far firms' choices differed from those that would have made optimal use of the nation's resources. According to the findings, even if the manager were to choose the technology that entailed the least cost to the firm, a more labor-intensive technology was generally more appropriate from the national point of view. The difference in optima resulted, of course, from the fact that the actual costs of factors of production did not always match their opportunity cost to the economy.

Williams, in the final study on technology choice reported in this book (Chapter 7), was the only author to concentrate on the technology choices of state-owned firms, although some others had included a few such firms in their samples. These other authors had noted that state-owned firms tended to choose capital-intensive techniques, presumably because of their insulation from competition. Williams concluded that the behavior of state enterprises, at least in Tanzania, has been more complicated than the other studies had suggested. The state-owned firms have not consistently chosen capital-intensive techniques, and their choices have been based on bureaucratic process rather than on economic criteria. At times the critical element in choice of technique was the speed with which a project could be implemented, while at other times the critical influence on the state firms came from external sources, such as foreign aid agencies and international financial institutions.

The studies included in this book provide quite compelling evidence that the rather simple economic model widely held before the availability of these studies was not adequate for explaining managers' choices of technology. In some cases, the model's assumptions about the market in which the manager operates were not appropriate. The simple model, for example, assumed that the manager has perfect information, when in fact *lack of information* is an important determinant of choice. Furthermore, the model assumed competitive markets, but many choices are actually made in an oligopolistic environment that results from the ability of some firms to differentiate their products or from the protection provided by governments. Nevertheless, the simple economic model would probably still be adequate if managers focused their attention on cutting manufacturing costs to the bone. It appears, however, that this is not always the sole— or even the principal—objective at which management decisions are aimed.

Moreover, the simple model does not reflect some of the trade-offs that often complicate the choice of technology. Production specialists have pointed out that the pursuit of lower production costs may reduce the flexibility of the plant. (This appears to be the case in the firms studied in this group of research projects.) In consideration of this and other trade-offs, managers appear to be willing to incur higher manufacturing costs in order to avoid certain risks, provided the market in which they operate is sufficiently uncompetitive that they are able to bear the additional costs associated with preferred techniques.

Of course, the simple model that is the favorite target of a number of these researchers is no longer one that commands the allegiance of all economists. In addition to the studies in this group, a number of other fine ones conducted elsewhere have already led to considerable change in the way the profession thinks about choice of technology.[5] To be sure, some of the findings from empirical analyses of managers' decisions still seem contradictory, but there is general agreement that managers do not behave in the ways that were once assumed to be the case.

If managers do indeed behave in the ways suggested by the research presented here, the implications for developing countries are profound. The usual policies to influence the kind of technology imported and its local adaptation are likely, in many cases, to have little impact. Conventional wisdom suggests that policies of "getting the prices right" for labor and capital will lead to socially optimal choices by managers, who, it was once widely assumed, always minimized costs and hence would select socially optimal technology if the factors of production were priced correctly in the economy. In other words, optimal technology would result if the cost of capital were held high enough to reflect its opportunity cost, and if labor costs were held down to their opportunity costs to the economy or subsidized so that the cost to the firm was no more than the opportunity cost.

To be sure, the results of the studies included here are consistent with the idea of the need to "get prices right," but they suggest that this step alone is generally not sufficient to induce the manager to make the choices that are best for society. For this reason, effective government policies are likely to include measures to encourage price competition, reduce the risks associated with labor redundancies, and so on. In cases where such policies are not feasible, governments might have to adjust prices to the point of heavily penalizing firms that use socially inappropriate technologies, by imposing tariffs or taxes on certain kinds of machinery, for example, or by taxing products on the basis of the technology used to produce them. Of course, the latter approach is feasible only when there are clearly identifiable classes of technology. Although the studies in this book differ somewhat in their recommendations for effective policies, they all agree that the policies suggested by old theory are unlikely to be adequate in the noncompetitive environments that are typical of most developing countries.

CHANNELS FOR TRANSFERRING TECHNOLOGY

Beyond the choice of technology itself, manufacturing firms face other choices when they transfer know-how for use in foreign plants, such as the channel to be used for transferring proprietary technology. Three mutually exclusive channels are available: sale of the know-how to an unrelated party, use of the technology in a facility partially owned by the technology owner, or use of the know-how in a facility wholly owned by the technology owner. Although a technology-licensing agreement may be used in all three channels, the first channel is often referred to as an "arm's-length license," or simply a "license"; the second channel, as a "joint venture"; and the third channel, as a "wholly owned subsidiary."

In the mid-1960s, there was widespread fear among government officials in host countries and among many scholars that multinational enterprises would consistently choose the wholly owned subsidiary route to transfer technology across borders. This channel would be the principal choice, it was thought, because the "bundling" of technology, capital, and management associated with the subsidiary would allow a firm to extract higher monopoly or oligopoly profits than would the use of alternative approaches that separate these elements. The fear was reinforced when executives of multinational enterprises expressed the view that, for technology owners, licensing was generally "an inadvisable alternative" to ownership.[6] If this was to be the typical pattern, governments of countries that imported technology would be frustrated in their desire for local ownership and control over the means of production inside their borders.

Studies of management behavior yielded surprising results. In 1966, one researcher hypothesized that a firm would have to generate a continuous stream of innovations if it were to be able to maintain control of production facilities in less developed countries.[7] As a given technology became older, it would be more readily available from more multinational enterprises. Since governments tend to prefer less ownership by foreign firms, some of the new entrants would bid for a place in foreign markets by offering know-how with less foreign ownership. As competition increased, firms would eventually make the know-how available through the licensing channel, with no foreign ownership. This hypothesis was supported, and others were tested, by Stobaugh's study of the petrochemical industry (Chapter 8), which used data on nearly all production facilities in the world that manufactured any one of nine petrochemicals and covered close to 600 technology transfers, about half of which were to plants located outside the U.S.

The firm's decision to transfer its manufacturing technology by licensing or by investing in a facility involves an evaluation of the benefits and costs to the enterprise of each approach. The results of this evaluation depend on the location of the facility, the competition faced by the firm, and certain of the firm's characteristics that are indicative of its resources and long-term strategies.

When international borders are crossed, the decision is more complex than when the transfer is within a country, and a firm is much more likely to choose the licensing channel. One reason for this is quite straightforward: licensing a firm abroad is less likely to create a new competitor for the technology owner than is licensing a firm at home.

Moreover, in comparison with using the technology investment in one's own foreign subsidiary, selling technology to a firm abroad requires less management skill and less knowledge of conditions abroad, it involves less risk, and it is more acceptable to some foreign governments.

Certain characteristics of the foreign country to which the technology is being transferred also affect the decision about the transfer channel. Although managers generally express a preference for ownership over licensing when the transfer is to a nation with a large market, the empirical evidence on petrochemicals indicates that they practice the opposite behavior. The high frequency of licensing in cases where the transfer is to a large market may be due to the pressure of indigenous firms in those countries, which are capable of purchasing and using the technology with little assistance from the supplying firm. Furthermore, the higher level of competition in large-market nations may make investment appear risky; direct investment can result in a substantial loss for the investing enterprise. Licensing, in contrast, almost always brings some return to the owner of the technology.

Just as the extent of competition influences the choice of technology, it also affects the choice of the channel through which technology is transferred. Stobaugh found that, when only a few petrochemical firms possessed competing technologies, most transfers were effected through wholly owned subsidiaries, but that, when many firms owned similar technology, the use of joint ventures or licensees was much more prevalent. The greater use of joint ventures or licensing arrangements may reflect the lower level of oligopoly profits available when competition increases. In this situation the firm providing the know-how may be willing to share the returns. Another factor may be that, in a competitive environment, the technology owner is simply not able to impose its will on an important country or firm.[8]

The type of competitor faced by an enterprise is also important in the choice of transfer channel. When engineering contractors own the technology, licensing is more prevalent than when the technology is held only by manufacturing firms. Indeed, when there are many owners of technology and some are engineering contractors, licensing is the dominant channel for international transfers.

Stobaugh also explored the relationship between the transfer decision and certain characteristics of the manufacturing firm that owned the technology. Large firms tended to establish their own subsidiaries because they were better able to accept the risk of a foreign investment than were small firms. Other things being equal, the nationality of the technology owner did not seem to be an important determinant of the channel through which know-how was transferred.

Telesio's study of foreign licensing by multinational enterprises (Chapter 9), the most comprehensive to date (as of 1983) on the subject, was based on data collected from in-depth interviews and from questionnaires covering managers of sixty-six U.S. and non-U.S. enterprises in a range of industries. The results confirmed the importance of competition and of the size of the firm in the decision about licensing or investing.

Telesio also argued that a firm's foreign manufacturing experience influenced its preference. Multinational enterprises that have relatively little experience with manufacturing in foreign markets will generally value the production capabilities of local companies highly and hence will be more likely to license than will experienced enterprises.

Certain strategies of firms are associated with arm's-length licensing as a method of transferring technology abroad. Highly diversified firms have to allocate company resources over many product lines in disparate markets, so some product lines may not get sufficient funds to support controlled investments in all foreign markets in which the firm wants to use its technology. Licensing provides an alternative to manufacturing in these markets. In addition, since each product line accounts for only a small percentage of sales and profits, a diversified firm is likely to be less concerned about the possible loss of control over a specific technology or market. For these reasons, highly diversified firms are more likely to be licensors than are less diversified firms. Moreover, firms that spend relatively more on research and development (R&D) are likely to have more opportunities for licensing, and hence are more likely to grant licenses, than firms that spend less on research and development.

The level of R&D expenditures can affect licensing activity in another way as well. For twenty-three multinationals in Telesio's sample of firms, licensing was not simply an alternative to investment but also a means of exchanging technology with other firms. Reciprocal licensing arrangements avoid redundant R&D in an industry and reduce the risk that any one firm will be left out of important technological innovations. Multinational enterprises that license for reciprocity are generally found in the pharmaceutical, chemical, and electrical industries, where on the average firms spend more on R&D as a percentage of sales than do firms in other industries.

Before Rafii completed his study of joint ventures and technology transfer in Iran (Chapter 10), government policy makers had little means of knowing whether the economic benefits to the country in which a production facility is located vary systematically with the degree of foreign ownership and control over the plant. Some authors had argued that ownership of subsidiaries by multinational enterprises provided net benefits far exceeding those associated with licensing. If this were true, host governments might want to adopt an open-door policy toward investment by multinationals; but if licenses provided more advantages, as many other authors had contended, governments might rather encourage licensing as a way of obtaining technology from abroad.

Rafii's study of thirty-five joint manufacturing ventures in Iran suggested that a higher degree of foreign ownership and control does impose higher costs on the host economy. Facilities with a large fraction of foreign ownership and control tended to have less domestic content in the final products, a lower share of indigenous management, and higher prices on imported equipment, raw materials, components, and technology than did their less controlled counterparts. In some of the foreign-owned and controlled operations, Rafii found very substantial overpricing of imported goods, with transfer prices running more than twice the free-market price and accounting for a substantial portion of the sales price of the final product. On the other hand, the same research uncovered evidence, albeit weaker than that for other findings, that a greater degree of foreign ownership and control was associated with greater manufacturing efficiency, as measured by price competitiveness with imported goods and the costs of domestic resources used in production.

The three studies on ownership and licensing suggest that a multinational enterprise often can extract a higher price from the local economy by investing in a manufacturing facility rather than by licensing know-how to another firm, typically a local one. But no one channel is always best from the firm's viewpoint: the country in which the facility is located, the competition faced by the firm, and certain characteristics of the firm all affected the preferences of the enterprises that have been studied. Clearly, there is no evidence that any one channel is ideal for all managers in all situations.

From the national viewpoint, the government official must consider the sources from which technology might be available and weigh the benefits of the potentially greater efficiency of foreign investment against the potentially greater costs, which, according to Rafii, are lower levels of local integration and local management opportunities and higher prices for imported goods and technology. A rigid policy with respect to foreign ownership is probably not in the best interests of the firm or the nation.

Like the results of the research on choice of technology, the findings of researchers concerned with channels of technology transfer are being merged, to some extent, with the theories of economics. Oliver Williamson's work on the internalization of transactions by firms has provided some of the basis for integrating the empirical discoveries about management preferences with economic theory.[9] But managers' preferences do not always determine the outcome. When industries are oligopolistic, there is room for bargaining between the host government and the foreign supplier of technology. Only analysis of relative bargaining power seems capable of predicting whether managers can follow their preferences with respect to channels for the transfer of technology.

MANAGEMENT OF OVERSEAS R&D AND MANUFACTURING

Technology is not always transferred intact from the domestic operations of a multinational enterprise to its foreign subsidiaries; technology may be modified, and new technology generated, by the multinational in its foreign subsidiaries. The location and function of research and development activity is important to multinational firms and to nations, which value the economic benefits derived from the skills, incomes, and improved balance of payments associated with local R&D activities. Based on the first systematic study of the location of R&D facilities by multinationals, Ronstadt describes fifty-five R&D investments operated abroad by seven U.S. enterprises (Chapter 11). Four different types of foreign R&D units are identified, and their evolution is mapped over time.

According to the study, most R&D units abroad are established as technical service laboratories to help transfer U.S. technology efficiently before the product or process technology has stabilized. Ronstadt calls these laboratories "technology transfer units." Such units also provide related technical services to foreign customers. Eventually, when the stream of innovations from the United States is perceived to be declining to such an extent that it is insufficient for the growth needs of the foreign business, a substantial increase in R&D efforts is made by some of the foreign subsidiaries so that they can generate improved and new processes and products for their own markets. In Ronstadt's terminology, this transformation creates "indigenous technology units."

One of the firms studied represents an exception to the usual pattern. Some of its foreign R&D units were established with the original mission to develop new products and processes for application in major world markets of the enterprise—"global technology units" in Ronstadt's parlance. These operations were established only after the parent had substantial production and marketing operations abroad, when one product line was being developed for worldwide markets, and when the already existing R&D facilities had reached a capacity limit set by the firm.

These three types of R&D units seem to have been generally successful. In contrast, a fourth type of R&D facility, created to generate new technology of a long-term or exploratory nature expressly for the corporate parent, failed in three of four instances. These "corporate technology units" were created when the parent decided to recruit top scientists located abroad who were unwilling to move to the United States. The units that failed did so because they did not receive a sufficient allocation of R&D resources for exploratory research within a focused technological area.

Once a firm decides to transfer technology to a manufacturing facility abroad, management must decide who in the enterprise will make the manufacturing decisions: the parent (or its agents in a regional headquarters) or the subsidiary. On the one hand, the essence of a multinational enterprise is that a vast network of manufacturing facilities is operated as a system, thereby improving upon the results that could be obtained if each unit acted independently. On the other hand, the headquarters cannot make all of the thousands of decisions that must be made in order for its different plants to function, nor can it be as sensitive as a local manager to the demands of the local market or the host government.

De Bodinat reports on an examination of the location of influence over manufacturing decisions in thirty-three U.S. and European multinational enterprises (Chapter 12). Three purposes of influence were identified: to transfer knowledge, to reduce slack in operations, and to coordinate the activities of different units in the multinational system. De Bodinat studied not only influence achieved through direct decision making at headquarters, but also two indirect methods of influence: the use of control systems, such as budget processes, and the role of corporate acculturation.

De Bodinat found that the variable with the strongest impact on influence was the degree of interdependence of the different operating units in the corporate system. A high level of interdependence causes headquarters to exert a direct influence on all three levels of manufacturing policy: major decisions, such as the selection of a process or the decision to add capacity; standards and procedures, covering such items as quality and cost; and day-to-day management of such matters as inventory and production scheduling. Moreover, parent companies exerted a strong indirect influence through control systems in cases where interdependence was high. A second variable, technology, was also found to have a strong influence, but primarily on major manufacturing decisions, for which a high degree of technological complexity results in centralized decisions at the corporate level. De Bodinat refers to interdependence and technical complexity as "push" variables because they encourage the parent to exert a high level of influence.

Two other variables also were found to be important—the uncertainty of the manufacturing task and the heterogeneity of the manufacturing system, which de Bodinat calls "barrier" variables. When they are high, the parent is unable to exert a high level of influence.

Obviously, these four variables do not exhaust explanations of influence of the parent enterprise. Other factors, such as performance of the unit and its cultural or geographical distance from the center, are also likely to be important.

The managerial issues covered in these last two studies are the most difficult to integrate into broader theory, yet they lie at the heart of many of the concerns of policy makers. The questions of where R&D is done and who maintains control over manufacturing facilities have both economic and political significance. An understanding of how multinational firms behave with respect to these two areas is essential if government officials—as well as local business managers—are to make sensible policies regarding the acquisition of foreign technology.

CONCLUSION

The eleven research projects described in this book go a long way toward improving our understanding of how business firms choose technology for international transfers and how they make management decisions with respect to their foreign manufacturing plants. Although the studies draw on the concepts of the economist, the empirical data collected by the researchers demonstrate the necessity of understanding more about management behavior than can be incorporated in neat models; the usefulness of concepts drawn from studies of the behavior of humans in complex organizations is also apparent.

The results of studies of the kind presented in this volume are in some ways less satisfying than those of the narrow economic models; but in exchange for elegance, their approaches make for great gains in descriptive power. Theoreticians in the future will be able to make even further progress by incorporating into new models the kinds of observations made by field researchers. For the time being, though, the kinds of conceptualizations offered in these studies appear to be the best available for managers and government officials struggling to make choices in the best interests of their firms or countries.

Part 1

Choice of Technology

2

Selection of Production Processes by U.S.-Based Multinational Enterprises

WAYNE A. YEOMAN

Multinational enterprises face wide differences in factor costs and scale requirements around the world, so it is not surprising that they often modify their domestic production methods for use in their foreign manufacturing subsidiaries. This chapter describes the kinds of changes that some firms have made and tries to determine the reasons for these adaptive patterns.

The empirical work suggests that conventional economic theory is inadequate to explain the behavior of multinational firms. Two variables outside the traditional models—cross-elasticity of

Adapted by the editors from the author's "Selection of Production Processes for the Manufacturing Subsidiaries of U.S.-Based Multinational Corporations," D.B.A. dissertation, Harvard Business School, 1968.

demand and the relative importance of manufacturing costs—must be considered if we are to understand how business managers determine the mix of labor and fixed capital in their production facilities.

CONVENTIONAL THEORY

The conventional theory (as of 1968) concerning choice of technology begins with a set of simple assumptions:

1. Firms behave rationally to maximize their self-interest.
2. Firms exist in a perfectly competitive market in which both buyer and seller are "price takers," in that neither can control price.
3. Economic units are small and numerous.
4. Products are undifferentiated.
5. Resources are mobile within the economy.
6. There are no barriers to entry or exit.
7. All firms possess perfect knowledge.
8. Resources are very versatile, that is, useful for the production of a wide range of goods and services.
9. And production factors are homogeneous and comparable across countries.

The relative prices of labor and capital determine which combination of the two will result in the lowest manufacturing costs. According to the assumptions of traditional theory—such as profit-maximizing behavior with perfect knowledge and perfect competition—a manager will always select the production process with the lowest manufacturing costs. Because factor costs vary by country, the optimal ratio of labor to capital, and hence the process chosen, will vary by country—especially with respect to developed versus developing countries. (Refer to the Appendix to this chapter for a graphical presentation of the traditional production theory.)

Many questions arise even within the framework of this traditional model. Perhaps the most important is whether a wide enough range of technologies is actually available to enable the manager to use various mixes of capital and labor. A second important question is whether the factors of production are homogeneous and comparable across countries. It is possible that wage differences reflect only differences in the quality of labor.

There is considerable evidence that one cannot simply set aside the assumption that labor can be substituted for capital, and vice versa, on the premise that modern technology dictates fixed proportions of labor and capital in the production process. Germany, Great Britain, and Japan, for instance, used different production processes during the war to produce similar products.[1] Other evidence comes from microeconomic studies of production processes.[2]

This study provides additional evidence that the mix of labor and capital is not rigidly set by modern production technology. A wide range of techniques was encountered in the operations of thirteen firms in three industries: pharmaceuticals; farm machinery, construction, mining, and materials-handling machinery and equipment (hereinafter, "machinery and equipment"); and household appliances. Only in the machinery and equipment industry is it difficult to find dramatic differences in factor proportions across the plants, and the field research suggests that even here the proportions of capital and labor need not be fixed. Company engineers of one of the machinery and equipment firms designed and installed an extremely labor-intensive tractor assembly line in a small plant in South Africa. The line handles 20 percent more volume than a similar line in Mexico and was built for approximately one-third the cost.

The study also found ample evidence of wage differences among countries ("real" wage differences in the sense that wages are different after correcting for differences in labor efficiency and for indirect labor costs). The capital-intensive firms in the study reported little difference in the productivity of labor in the various countries.[3] For labor-intensive operations, on the other hand, labor in the developing countries is less productive than in the advanced countries, but the difference is not nearly enough to offset the wage differentials. Moreover, the wage differences are not offset by differences in indirect costs associated with the labor.

ADAPTIVE PATTERNS IN THE SAMPLE FIRMS

Although conditions for the sample firms met certain assumptions of traditional theory—that firms can choose from a wide range of techniques and that there are real differences in factor costs from country to country—the conventional model does not explain the actual choices made by the companies in this study, whose observed behavior varies from industry to industry.

Pharmaceuticals

In three of the four drug companies studied (Companies A_1, A_2, and A_3), plant and process design for facilities abroad was a joint responsibility of research personnel and engineering staff located in the United States. Interviews with the key design people for these three companies pointed to two important conclusions: relative factor costs around the world had not been a consideration in their selection of manufacturing processes; and these firms evidenced little or no interest in the possibilities for factor substitution abroad.

An analysis of the capital-labor ratios in a range of plants operated by these companies confirms the managers' statements. The ratios must be interpreted with some care, however. The capital figures used represent gross, rather than net, fixed capital. Figures net of depreciation could be derived, but the results might be misleading because of widely differing depreciation rates across countries. Fortunately, almost all of the plants included in the analysis were part of the great wave of expansion that had come in the half-dozen years prior to the study, so they were of approximately the same age.

Differences in construction and equipment costs around the globe also complicate the interpretation of fixed capital investment figures for firms in different countries. This problem is relatively minor, however, because these variations are not great, and to some extent they are mutually offsetting. Typically, construction costs are lower and equipment costs are somewhat higher in the less developed areas than in the developed countries. No adjustments have been made in the investment figures to compensate for these differences in construction and equipment costs.

If comparisons of capital-labor ratios are to be meaningful, capital must be utilized at approximately the same rate in each plant. The comparisons in this analysis are based upon single-shift operations at design capacity.

The labor figures shown for each facility include the following personnel: manufacturing plant labor plus those engaged in manufacturing support, warehousing, and general administration. Although there are minor differences among the various corporations in the classification of employees, the data are consistent within each firm.

The combinations of capital and labor employed in twelve manufacturing facilities of drug company A_1 are shown in table 2–1. While the products of these twelve plants were not identical, the operations in each were quite similar: pharmaceutical formulation, packaging, and warehousing. The capital intensity was remarkably uniform, varying from a low of about $8,000 to a high of close to $16,000 per employee.

Table 2–2 presents the capital-labor ratios of Company A_2's foreign facilities with roughly comparable operations. These data likewise suggest very little concern for local factor availability. The most capital-intensive manufacturing units (except one in Australia) were located in less developed countries. Any adjustment for labor productivity would, of course, increase the ratios for these Latin American plants. Data from Company A_3 show a similar lack of difference in factor costs affecting the selection of manufacturing processes.

Company A_4 is an exception to the pattern found in the other

TABLE 2–1

CAPITAL-LABOR RATIOS—COMPANY A_1

PLANT LOCATION	GROSS FIXED CAPITAL ($000)	LABOR FORCE	CAPITAL-LABOR RATIO (dollars per employee)
Venezuela	$1,164	73	$15,950
Brazil	3,144	247	12,700
Netherlands	2,827	225	12,550
Mexico	1,367	124	11,000
India	4,223	386	10,950
Canada	1,523	143	10,650
Thailand	928	88	10,550
United Kingdom	3,479	346	10,050
Peru	554	57	9,700
Italy	511	53	9,600
Argentina	1,128	131	8,600
Philippines	453	56	8,100

TABLE 2–2

CAPITAL-LABOR RATIOS—COMPANY A_2

PLANT LOCATION	GROSS FIXED CAPITAL ($000)	LABOR FORCE	CAPITAL-LABOR RATIO (dollars per employee)
Colombia	$2,939	148	$19,850
Australia	5,574	290	19,200
Argentina	5,016	364	13,800
Venezuela	1,364	100	13,650
Brazil	2,264	174	13,000
Canada	3,104	290	10,700
France	997	106	9,400

three drug firms. Local managers in this corporation had a significant say in the design of their manufacturing facilities, and, as the data in table 2–3 indicate, in many cases the managers exercised the option to substitute labor for capital.

Machinery and Equipment

Of the six companies in the sample drawn from this industrial classification, three (B_1, B_2, and B_3) manufactured farm machinery

TABLE 2–3

CAPITAL-LABOR RATIOS—COMPANY A_4

PLANT LOCATION	GROSS FIXED CAPITAL ($000)	LABOR FORCE	CAPITAL-LABOR RATIO (dollars per employee)
Japan	$9,890	720	$13,750
Germany	4,758	508	9,350
Canada	1,842	299	6,150
Portugal	1,000	176	5,700
Iran	1,100	240	4,600
Mexico	1,911	424	4,500
India	3,864	1,519	2,550

and construction equipment, one (B₄) manufactured mining equipment, and the other two (B₅ and B₆) produced materials-handling equipment. In most of these corporations, the production executives and the plant and process design engineers stated that relative factor costs around the world played little or no part in the design of their plants. There were exceptions to this pattern, however.

Company B₁'s manufacturing plants abroad were reported to be carbon copies of its U.S. facilities. The variations in the capitalization of the four construction equipment plants were relatively small (see table 2–4) and resulted for the most part from differences in the product lines manufactured in the various facilities.

The story on factor substitution was much the same with Company B₂. Engineers in the international division of this firm reported that its domestic standards had been copied completely in both its major and its smaller foreign plants.

Manufacturing plants in Company B₂ were not completely comparable: its range of products was quite broad and its production plants were rather specialized in their output. Table 2–5 includes capital-labor ratios for two of the firm's U.S. plants, to indicate the characteristic degree of differences in capitalization between the tractor and the farm implement product lines and to provide a basis for comparing B₂'s U.S. and foreign facilities.

The data suggest that Company B₂ has been insensitive to factor-cost differences in the design of its manufacturing processes. While the foreign capital-labor ratios in table 2–5 are below the U.S. benchmarks, the differences for similar facilities are so small that they may reflect nothing more than intercountry differentials in construction

TABLE 2–4

Capital-Labor Ratios—Company B₁

Plant Location	Gross Fixed Capital ($000)	Labor Force	Capital-Labor Ratio (dollars per employee)
France	$26,422	1,425	$18,550
Brazil	11,534	742	15,550
Australia	8,147	664	12,250
United Kingdom	38,030	3,752	10,100

TABLE 2–5

CAPITAL-LABOR RATIOS—COMPANY B_2

PRODUCT LINE AND PLANT LOCATION	GROSS FIXED CAPITAL ($000)	LABOR FORCE	CAPITAL-LABOR RATIO (dollars per employee)
Tractors			
United States	$54,450	4,291	$12,700
France	28,405	2,496	11,400
Germany	32,570	3,284	9,900
Mexico	3,336	743	4,500[a]
Farm Implements			
United States	24,694	2,718	9,100[b]
France	6,329	740	8,550[b]
France	18,290	3,126	5,850[c]
Sweden	3,584	591	6,050[c]
Construction Equipment			
Germany	6,259	678	9,250

[a]This is essentially a large-scale job-shop operation. While enough tractor parts are manufactured in the facility to satisfy the Mexican government's value-added requirements, the items that are most difficult to produce (engines, crankshafts, differentials, special gear trains, etc.) are imported. The plant and production line were designed by the corporate headquarters engineering staff; U.S. design criteria were used throughout.
[b]These plants produce light farm implements—harvesting and hay machines—and also include a twine mill. U.S. facilities of this type are capitalized at approxiately the same level. The twine mill in France is reported to be more modern than B_2's largest U.S. twine mill.
[c]These facilities produce tillage tools.

and equipment costs or minor variations in product line. The low capital-labor ratio for Mexico, for example, appears to be accounted for by the different nature of the operation there compared with elsewhere.

The attitude toward factor substitution was somewhat different in the case of Company B_3. The senior manufacturing executives in this farm machinery and construction equipment firm expressed considerable interest in the possibilities for factor substitution. B_3's very labor-intensive assembly line in South Africa takes maximal advantage of the cheap labor that is available locally. The South African plant handles 20 percent more volume than B_3's Mexican facility and was tooled at one-third the cost.

Apart from its South African plant, however, B_3's foreign

facilities have been designed to U.S. specifications. A company decision points up the similarities that are typical among B_3's major
manufacturing plants. The company had approved both a sizable
expansion of a domestic plant and the construction of a new facility in
Europe to produce the same piece of equipment. After the production
line and support facilities for the U.S. addition were designed, the
order for equipment was simply doubled and the same technology was
installed in Europe. In B_3's other major foreign facilities the differences from U.S. practices are minor and have not resulted from any
conscious effort to adapt to local factor availability.

Again except for the South African plant, the capitalization of
B_3's smaller manufacturing and assembly units abroad runs about
$5,000 per employee, a figure that is very close to the capitalization of
Company B_2's plants of comparable size. The design of these minor
facilities was handled through B_3's corporate headquarters and was
done by engineers from the firm's low-volume implement operations.

Corporate executives of Company B_4, a firm that manufactured a
broad line of mining equipment, indicated that relative factor prices
had little if anything to do with the design of production processes for
its overseas facilities. Except for adjustments to reflect scale of operations, the manufacturing methods used in B_4's plants in Mexico, Peru,
and South Africa were reported to be very similar to those used in
the United States, Canada, Great Britain, France, Belgium, and
Australia.

Company B_4 would not release detailed information on its individual manufacturing plants abroad but did make available certain
data on the combined operations of its international subsidiaries.
These data—while they obscure any differences among the various
facilities—confirm that, on the average, B_4's foreign operations were
quite capital intensive. The average annual value added for each
manufacturing employee throughout the foreign plants in this company was very high—$13,160. For direct manufacturing labor alone,
the average figure in the international subsidiaries was $24,700.
Clearly, this kind of performance would not be possible unless B_4's
production employees were supported by a very heavy investment in
capital equipment.

The fifth firm in this category, Company B_5, produced a broad
range of materials-handling and materials-processing equipment. Rel-

atively little detailed data on the foreign manufacturing plants in this system were available at the corporate headquarters, but interviews with B_5's international production executives suggested that production facilities abroad were similar to U.S. plants producing the same volume of output. The company's Mexican plant seemed a valid illustration of corporate thinking on the design and operation of production units abroad. One of the newest facilities in the system, it had been designed by a team of engineers from the company's largest plant. Aside from certain adjustments necessitated by scale differences, equipment and layout specifications were identical to those of the U.S. production line.

Company B_6, which manufactured elevators and escalators, generally produced at low volume and used a relatively simple manufacturing process. This company's interest in the possibilities for factor substitution far exceeded that of any of the other five firms in this category. The head of B_6's industrial engineering staff summarized the company's position as follows (in an interview conducted in the mid-1960s):

> The choice of equipment which is considered appropriate may vary because of local conditions unrelated to the equipment itself. For instance, hourly wages in the United States and Canada are extremely high and thus time saving is of paramount importance. For this reason, the use of more automatic machinery can be justified in the United States and Canada than in Europe, where wages are only one-third as high. Similarly, European wages may be five times higher than those in Latin America or India, and for this reason semiautomatic equipment may be appropriate in Europe but not in the other areas. Thus, to perform exactly the same function we may use completely automatic equipment in the United States, semiautomatic equipment in Europe, and conventional hand-operated equipment in other areas. The criterion which is used is the relationship of capital cost to the local wage rates modified by the productive efficiency of local labor. Thus, we may have three entirely different solutions for the same problem, yet all of the solutions will be logical.

Because figures on the total investment in fixed assets at each plant could not be made available, relative capitalization levels in

Company B_6's various production units were inferred from an analysis of value added per employee. Table 2–6 compares indices of the value added in manufacture by an average production employee in nine of the company's plants during 1964. The labor units used in the analysis were adjusted for differences in worker efficiency. With the effect of variations in the quality of labor eliminated from the data, capitalization per employee becomes the important determinant of any differences in the value of output per worker. Thus, the indices in table 2–6 reflect the plants' levels of capitalization. The data support the company's contention that labor was substituted for capital in its production facilities where relative factor prices favored such a response. B_6's manufacturing units in high-wage areas were capitalized at much higher levels than those in low-wage countries.

Household Appliances

The three companies producing household appliances had significantly adjusted their domestic production methods for the manufacturing facilities they built around the world. Two of the companies, C_1 and C_2, appeared to be relatively satisfied with the design of their foreign production plants. The third firm, Company C_3, reported that it had gone too far in the substitution of labor for capital; it expected to increase the capitalization of future plants by at least 25 percent.

TABLE 2–6

INDICES OF VALUE ADDED PER LABOR UNIT—COMPANY B_6

PLANT LOCATION	INDEX NUMBER
Australia	100
United Kingdom	89
Italy	87
France	68
Mexico	68
South Africa	65
Argentina	45
India	40
Brazil	21

TABLE 2–7

CAPITAL-LABOR RATIOS—COMPANY C_1

PLANT LOCATION	GROSS FIXED CAPITAL ($000)	LABOR FORCE	CAPITAL-LABOR RATIO (dollars per employee)
United States	$32,880	2,249	$14,600[a]
France	8,830	718	12,300
United Kingdom	58,276	7,122	8,200
Canada	1,975	285	6,950
Mexico	995	138	7,200
Brazil	1,040	147	7,050
Australia	2,868	578	4,950

[a]This is C_1's major U.S. production facility.

The extent of factor substitution in the manufacturing units of Company C_1 is reflected in table 2–7. The company's U.S. and French plants were designed by American engineers, while its other production facilities were handled by British engineers. The company's British design team showed a preference for labor even in the U.K. plant, a facility that was almost twice as large as any other in the system.

Company C_2 had substituted labor for capital in much the same way as did C_1 (see table 2–8).

TABLE 2–8

CAPITAL-LABOR RATIOS—COMPANY C_2

PLANT LOCATION	GROSS FIXED CAPITAL ($000)	LABOR FORCE	CAPITAL-LABOR RATIO (dollars per employee)
United States	$ 9,462	597	$15,850[a]
Canada	6,184	542	11,400
Australia	15,313	2,204	6,950
New Zealand	718	127	5,650
Mexico	1,296	273	4,750

[a]This is one of several plants operated by C_2 in the United States; its capitalization is typical of the company's domestic facilities.

TABLE 2–9

CAPITAL-LABOR RATIOS—COMPANY C_3

PLANT LOCATION	NET FIXED CAPITAL[a] ($000)	LABOR FORCE	CAPITAL-LABOR RATIO (dollars per employee)
United States	$8,000	1,932	$4,150
Germany	5,978	2,056	2,900
Japan	1,121	1,012	1,100
Brazil	1,086	1,734	650

[a] All of C_3's plants had been operating for a long period of time. In this company, the net asset figures across the system give an accurate comparison of the capital available to the work force in each of the various plants.

Production methods used in C_3's four major manufacturing plants differed substantially (see table 2–9). On average, for example, C_3's U.S. workers were provided more than six times as much capital as the company's Brazilian workers. The disparity between U.S. and Japanese operations was also great, with Japanese workers receiving one-quarter the capital support of their U.S. counterparts.

Company C_3 had had continuing problems in maintaining control of the manufacturing process in certain of its foreign plants, part of which management attributed to an excessive substitution of labor for capital in these operations.

EXPLAINING THE EVIDENCE

Contrary to the predictions of conventional production theory, the evidence indicates that the sample companies responded to differences in relative factor costs in very different ways. In the pharmaceutical and the machinery and equipment industries, most firms showed little inclination to adjust processes as they built facilities abroad. All the small appliance companies, by contrast, adapted their processes to local conditions (see table 2–10).

The metal-working firms engaged in very heavy manufacturing tended to adapt production operations somewhat differently from those engaged in light manufacture. In general, there was much less response to factor price changes among the heavy equipment produc-

TABLE 2–10

DEGREE TO WHICH COMPANIES ADJUSTED PRODUCTION PROCESS TO LOCAL FACTOR
COSTS

INDUSTRY	DEGREE OF ADJUSTMENT	
	Low	High
Pharmaceuticals	A_1, A_2, A_3	A_4
Machinery and equipment	$B_1, B_2, B_3{}^a, B_4, B_5$	B_6
Household appliances		C_1, C_2, C_3

[a]Company B_3 had successfully adjusted the technology used in one of its plants, in South Africa, in order to take advantage of low-cost labor.

ers than among the appliance manufacturers. This pattern is consistent with Gerard Boon's laboratory research, which indicates that, the larger the size of the work piece and the higher the precision requirements in production, the more the choice of technology is restricted.[4]

These generalizations, while helpful, fall short of explaining the adaptive patterns observed. In fact, it appears that an individual firm's choice of manufacturing methods may depend more on variables outside the production process than on any associated directly with it.

Two factors stand out as particularly important. The first is the firm's estimate of the nature of individual demand curves in foreign markets. Although executives discussed requirements for quality standards, product reputation, and keeping faith with old and trusted customers, these expressed concerns seem to have been code words for the cross-elasticity of demand of the products in question (that is, how sensitive the sales volume of a product is to the price of that product relative to the price of competing products).[5] The countries with cheap labor were, in many cases, the very countries where the firms had been granted a monopoly position. Under these conditions, product cross-elasticities are extremely low; as a result, the incentives to identify the least-cost production method are not great.

A second factor that seems to influence the firm's decision on manufacturing processes is the relative importance of production in the total effort expended in bringing products to market. A company that attributes its success to its marketing skills, for example, typically directs the bulk of its resources to this function. It is quite unlikely that such a company would commit managerial talents and funds to the development of a wide range of production processes, each designed to optimize a different combination of factor prices.[6] A gross measure of this variable is the fraction of the price of the product accounted for by the firm's manufacturing labor (direct and indirect) and fixed capital costs in the United States. The higher the total of plant labor and depreciation costs as a fraction of selling price, the greater the potential for profit from exploiting any changes in the relative prices of these factors. The empirical evidence indicates that the firms responded to this variable as one might have expected: where labor and depreciation costs constituted a small fraction of the selling price in this country, interest in developing new production methods for foreign plants was slight; as this percentage increased, the pressure for factor substitution in overseas facilities also increased.

To appreciate the effects of the two variables not accounted for in conventional production theory, it may be useful to consider the two polar cases, as shown schematically in figure 2–1. Firm A, with low cross-elasticity of demand and relatively insignificant manufacturing costs, experiences little pressure to change its U.S. production processes as it goes abroad. The corporate strategy of such a firm typically emphasizes maintaining a low cross-elasticity of demand for the product through product quality and market image. As a result, the firm is discouraged from accepting the cost and uncertainty of searching for alternative production methods. If production costs are higher than they might have been with factor combinations more appropriate to local conditions, the difference can be recouped through higher prices. The decision by Firm A to use its U.S. manufacturing methods abroad is further supported by the fact that labor and fixed capital costs are a small percentage of the selling price of the product. The potential cost savings of substituting labor for capital in overseas production facilities are judged to be limited.[7] They at least are not likely to be worth the associated cost and risks, particularly the possible downgrading of quality.

As one moves up and to the right in the grid, the pressures for adapting U.S. production processes to the structure of local factor costs increase. If the technology employed by Firm B permits substitution of labor for capital, one can be reasonably sure that B's production processes around the world will reflect the local prices of these two factors. Failure to adjust will result in higher costs than those of

Fig. 2–1. Schematic diagram of the effects of two variables that induce companies to adapt production processes to factor costs in different countries.

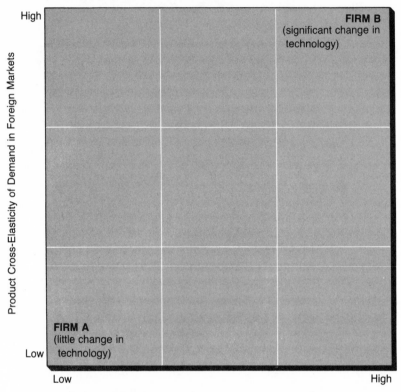

Costs of Manufacturing Labor and Depreciation as a Percentage
of Producer's Selling Price in the United States

competitors. But with high elasticity of demand, the company must meet competitors' prices or lose its market; it cannot adjust product price to cover unnecessary manufacturing costs. Moreover, the potential for sizable savings from factor substitution in the production process represents an additional incentive for Firm B to change methods. The greater the price cross-elasticities of demand and the larger the share of product price represented by labor and depreciation costs, the more likely that business behavior will approach the solutions suggested by conventional production theory. The data obtained during the interviews bear out the importance of these two variables.

Manufacturing Labor and Capital Costs

The potential savings from optimal production processes varied widely across the companies in the sample.

The very simple production processes used in the foreign manufacturing facilities of the drug firms (conversion of bulk chemicals to various dosage forms, packaging, labeling, etc.) are probably typical of an important class of plants abroad. The distribution of costs among research and development, production, and marketing appears to have a very significant effect on the firms' responses to various sets of factor prices.

While production personnel from the drug companies in the sample agreed that the manufacturing labor and depreciation costs together accounted for less than 5 percent of product selling price, the companies were reluctant to release any additional breakdown of costs. Considerable detail on the cost structure of the drug industry is available, however, from the 1959 Kefauver Committee Hearings on administered prices, which studied the subcontractors who prepare bulk drugs for sale to retailers.

The information in table 2–11, taken from the hearings, confirms that, in production operations of this type, capital and labor inputs are a minor part of total manufacturing costs and a very small part of selling price. In fact, the tableting and bottling costs listed in the table overstate the capital and labor elements of production costs. Since

TABLE 2–11

CONTRACT PROCESSING CHARGES—STEROID HORMONES[a] (PER THOUSAND TABLETS)

Bulk powder	$ 11.85
Allowance for wastage (5%)	.62
Tableting charge	2.00
Bottling charge (10 bottles of 100 tablets each)	1.20
Total production cost	$ 15.67
Selling price to druggists	$179.00

SOURCE: U.S. Congress, Senate Committee on the Judiciary, *Hearings Before the Subcommittee on Antitrust and Monopoly, Part 14, Administrative Prices in the Drug Industry,* 86th Cong., December 1959.
[a]The charges shown are for lots of 100,000 tablets. For lots of one million tablets, the tableting charge is reduced to $1.25 per thousand tablets.

these operations were performed by a contractor, the costs include the contractor's profit and the cost of bottles and shipping containers— perhaps as much as 10 to 15 percent of the total. If the several estimates submitted to the committee are averaged, the total of the contractor's capital and labor costs is between 1 and 2 percent of the drug manufacturer's selling price. This percentage would remain quite small even if the manufacturer drastically reduced the price of the tablets to the retailers. In this industry, research and development and distribution costs are the important ones.

Manufacturing efficiency is somewhat more important for the household appliance producers. The product lines and cost structures of two of the companies in the sample, C_1 and C_2, are quite similar. On the average, the cost of the capital and labor these firms expended in production made up between 5 and 10 percent of the selling price. C_3's product line was somewhat more labor intensive. At this company's main U.S. production facility, labor and depreciation accounted for almost 30 percent of the factory transfer price and 10.5 percent of the final selling price.

Least-cost production methods would appear to be even more critical to the firms that produce machinery and equipment. Companies B_1 through B_5 differed as to the distribution of their labor costs over direct and indirect categories, but there was very little variation in the total component of factory cost. The same can be said for capital consumption allowances. Capital and labor inputs together comprised about 15 percent of the selling price of these companies' products.

The distribution of manufacturing costs was significantly different for B_6 (a firm that produced elevators and escalators). The labor requirements of B_6's production processes were especially high, typically representing 60 percent of the firm's factory costs in its U.S. plants. The total of capital and labor costs typically represented 30 percent of the price for which B_6 sold its products in the United States.

As can be seen from figure 2–2, the relative importance of manufacturing capital and labor varied widely across the thirteen firms in the sample, accounting for less than 5 percent of product selling price in the four drug firms but roughly 30 percent in Company B_6. These differences appear to have influenced the companies' propensity to adjust their production processes as they built manufacturing facilities abroad.

Fig. 2–2.   Costs of manufacturing labor and depreciation as a percentage of producer's selling price in the United States for thirteen sample companies.

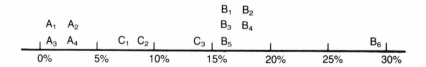

Cross-Elasticity of Demand

The estimation of a firm's demand cross-elasticities is very difficult. In several important respects, however, one can distinguish among the thirteen companies in the sample, estimating the relative power that each has over the pricing of its products.

There is strong evidence that drug firms were particularly free to set product prices within very broad limits. While the lack of price competition in the drug industry is due to several factors, the most important is the manufacturers' ability to differentiate their products. Drug firms have combined their efforts in new product development with large promotional programs directed toward persuading physicians to prescribe by brand name, a strategy that has met with great success.

There is even reason to believe that the multinational drug companies face lower cross-elasticities for their products in foreign markets than in the United States. In many countries producers are few, markets are protected, and demand is high and growing rapidly. A report on the operations of a sample of U.S. drug companies in India tends to support this proposition. Demand for drugs there was consistently outpacing supply, and "in India, high prestige is accorded to 'foreign' brand names. . . . Anything which has a foreign air about it, regardless of its source of origin, is accorded high status."[8] Profit rates for U.S. firms were several times higher than those for Indian companies selling in the same market. Some of the disparity in profitability is attributable to differences in product mix and efficiency, but there is also an "apparent tendency of U.S. companies to place higher selling prices on their products."[9]

This propensity of the drug companies to set higher prices in foreign markets was also noted by the Kefauver Committee. As indicated in table 2–12, prices charged by U.S. firms varied widely from country to country, with the variations bearing very little relation to cost differences; rather, they reflected the company's control over the price it could charge for its products.

It is difficult to justify any particular ranking of the four drug firms in the sample. Each of the companies had a large number of exclusive products with strong brand names. As a group, the drug firms deserve the four spaces at the head of the list for price-setting ability.

Similarly, it is relatively easy to place companies B_1, B_2, B_3, B_4, and B_5 in their proper place as a group but more difficult to justify a particular ordering of the five firms. Generally, the cross-elasticities in foreign markets for the principal products of these five firms were estimated to be quite low.

Bain's discussion and conclusions regarding the tractor and farm machinery industries are applicable to the three sample firms that produced construction and farm equipment (B_1, B_2, and B_3).[10] Much of the analysis is also relevant to B_4, a firm that produced mining equipment, and to B_5, a company that manufactured heavy and complex materials-handling equipment. The requirements for capital and technical skills in the manufacturing processes of B_4 and B_5 were quite similar to those of the other three companies. Bain's analysis leads him to make a number of predictions regarding the performance

TABLE 2–12

U.S. FIRMS' COMPARATIVE DOMESTIC AND FOREIGN PRICES TO DRUGGISTS (PREDNISONE AND PREDNISOLONE—BOTTLES OF 100 TABLETS)

United States	$17.90
Toronto, Canada	20.80
Rome, Italy	22.16
Colon, Panama	22.99
Sydney, Australia	24.00
Tokyo, Japan	27.78

SOURCE: U.S. Congress, Senate Committee on the Judiciary, *Hearings Before the Subcommittee on Antitrust and Monopoly, Part 14, Administrative Prices in the Drug Industry,* 86th Cong., December 1959, pp. 8065 and 8314.

to be expected of a typical firm producing tractors and/or other farm machinery in the United States. For our purposes the most important of his predictions are:

1. Prices substantially in excess of minimal costs.
2. Excess profits of very substantial size and generally at the highest levels to be observed.
3. Substantial selling costs for advertising, maintenance of distributive facilities, etc.[11]

We are interested, of course, in the behavior of specific firms from the sample and not with an industry average. In addition, we are concerned with predictions related to foreign markets rather than domestic ones.

In their U.S. operations, B_1, B_2, and B_3 were all doing extremely well. Their major product lines were well known and respected by U.S. consumers, and each had enjoyed rapid sales growth in the years just before this study. The production facilities operated by the three companies were uniformly modern and efficient; each was adding new manufacturing capacity at an impressive rate. Further, each company's domestic distribution and service systems were very strong. Clearly, then, there is every reason to believe that Bain's predictions were borne out by the U.S. operations of all three companies.

But the overseas story was different. B_1, the dominant firm abroad, had pushed its international operations very hard throughout the 1950s and 1960s, meeting with such great success that foreign sales accounted for approximately half of its business at the time of the study (1968). The company was producing at capacity in all of its foreign plants and was adding new manufacturing facilities as rapidly as possible. Because B_1's international dealer network was particularly strong, its customers were assured that parts and technical assistance would be available on short notice almost anywhere.

Although B_2 was also firmly established and doing well abroad, it consistently came off second best when in direct competition with B_1. Company B_3 was a late entry to the foreign market and had encountered serious problems with its international operations. It had lost money on its overseas operations throughout the 1960s, and losses by the international division in 1967 were greater than in 1966. B_3's managers pointed out the severity of price competition in foreign markets. The company's brand name was not yet well established abroad, nor was its product line differentiated to any significant degree. Cross-elasticities for B_3's products in overseas areas had been relatively high, certainly higher than those faced by B_1 and B_2. The firms should be ranked in the following order: B_1, B_2, and B_3.

For the B_4 and B_5 product lines, the cross-elasticities of demand in foreign markets were judged to be at least as low as those of B_1. Both companies produced specialty equipment, that is, machines for specific applications; both dominated their particular markets; and both possessed the engineering, manufacturing, and distribution resources required to gain and maintain a position of worldwide leadership.

The extent of B_4's dominance of its competitors is indicated by the fact that the company's biggest and most important division controlled 80 percent of the world market for its type of machinery. The firm had been in international markets for many years and had acquired an efficient and well-integrated manufacturing and distribution system.

Company B_5 was similarly the world's leading firm in its field. Much of the company's success had been built upon its research and development effort and its engineering staff. The company held a large number of patents on its various products and was the single-source supplier for a broad range of equipment.

The cross-elasticities for the product lines of the four remaining sample companies were all much higher than those discussed to this point. Our judgment is that B_6 (elevators and escalators) had slightly lower cross-elasticities than the three small-appliance companies.

Only B_6's marketing skills kept the cross-elasticity of its product line from being the highest of any in the sample. For several reasons, one might expect the elevator purchase decision to be made largely on the basis of price. The manufacturing technology is simple, economies of scale are relatively unimportant, and capital requirements for entry are small.[12] But, to a surprising degree, B_6 had successfully differentiated its products. In some markets, buyers were willing to spend heavily on such features as the decor of the cab. In these cases, price was not a critical consideration in making the sale.

Cross-elasticities in overseas markets were quite high for the products of all three of the small-appliance firms in the sample. This judgment had the unanimous support of the various officials of companies C_1, C_2, and C_3 who were questioned on the point.[13] Of the three companies, C_3 was perhaps the most concerned about this problem. The nature of C_3's product line was such that major design changes were infrequent. The firms in this segment of the industry competed with relatively standard items, and a number of Japanese companies had successfully entered the market.

It is easy to decide that the small-appliance firms as a group come last in our ranking of the sample. At the same time, there is little apparent basis for differentiating among the three companies. C_3's products seemed to face higher cross-elasticities than those of the other two firms, but the differences were quite small.

The Impact of the Two Variables

Figure 2–3 describes the thirteen sample firms in terms of the price elasticities of demand they faced and the role of manufacturing costs in their prices. Their positions in the matrix are consistent with the degree to which they had adapted their production techniques.

While scale requirements had dictated certain adjustments to U.S. equipment and methods, there is little evidence to suggest that drug companies A_1, A_2, and A_3 made any effort to exploit the differences in relative factor prices abroad. A_4 is the exception. Since it was the most decentralized of the drug firms in the sample, it is not surprising that it departed somewhat from the model.

A significant amount of substitution of labor for capital was found in the foreign plants of all three of the small-appliance firms. The high product cross-elasticities of demand in this industry had forced the companies to substitute cheap production factors for more expensive ones.

The production processes used in the foreign plants of all six of the machinery and equipment manufacturers seem consistent with the model. The foreign manufacturing plants of B_1, B_2, B_4, and B_5, each of which faced a low cross-elasticity of demand, were designed completely to U.S. standards. B_3's South African plant represented a first step toward what appears to be more rational use of labor and capital in the less developed areas. Our hypothesis is that this step was encouraged by the relatively higher cross-elasticities that B_3's products faced in foreign markets. No company in the sample appeared to be more sensitive than B_6 to the problem of factor availability and least-

Fig. 2–3. Adaptation by thirteen sample companies of production processes to factor costs in different countries, as a function of two explanatory variables.

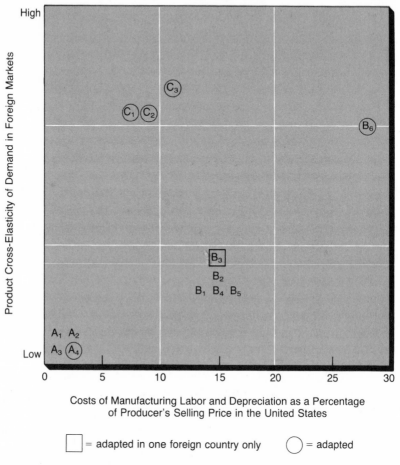

Costs of Manufacturing Labor and Depreciation as a Percentage of Producer's Selling Price in the United States

☐ = adapted in one foreign country only   ◯ = adapted

cost production methods. If our model is valid, B_6's conformance to production theory is not accidental but can be explained by the company's position in the grid.

To sum up, field research indicates that companies' choices of production processes for their overseas operations are significantly influenced by two factors not weighed in traditional production theory: the cross-elasticity of demand and the relative importance of manufacturing costs.

APPENDIX

A NOTE ON PRODUCTION THEORY

Labor and capital can be combined in many ways to produce a given volume of a product. The various efficient combinations can be represented as an isoquant, that is, a line of equal quantities, as shown in figure 2A–1. This isoquant shows the different combinations of capital and labor that can be used to produce 100 units of output. At any point on the curve, it is impossible to increase the output of the product without increasing the input of either labor or capital.

In figure 2A–2, the isoquant curve is superimposed on a set of isocost lines, each of which represents all of the combinations of labor and capital that yield a given level of total costs. In this example, the three different given levels of total costs are $25, $50, and $75. Thus, OB units of capital cost $50, OA units of labor cost $50, and any point on the line BA represents the mixture of capital and labor that would result in an expenditure of $50. The slope of the line is determined by the relative prices of units of capital and labor. The least-cost combination of factors will be found at the point where the isoquant is tangent to the lowest possible isocost—in this case, P_1, where 100 units of output would cost $50. In contrast, 100 units of output would cost $75 at P_2 or at P_3. Note that it is not possible to produce 100 units at any cost below $50.

Fig. 2A–1. Isoquant.

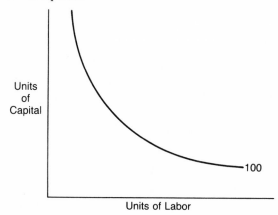

Fig. 2A–2.   Isoquant with isocosts.

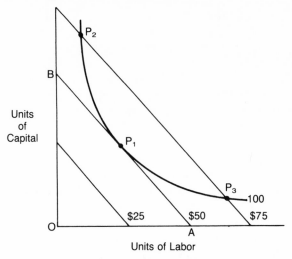

The slope of the isocosts will vary from country to country, reflecting the relative costs of labor and capital. Thus, as figure 2A-3 indicates, the optimal production technique will differ from country to country, while the total cost to produce 100 units remains at $50 in each country (in this example).

Fig. 2A–3.   Isoquant with isocosts of different countries.

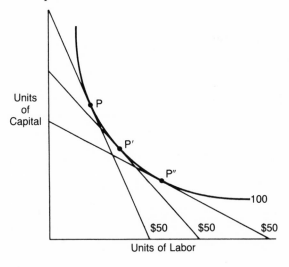

3

Economic Man and Engineering Man

LOUIS T. WELLS, JR.

One of the most perplexing problems that faced developing countries in the early 1970s was that of unemployment. Industrialization, it was generally thought, had not done its share to absorb labor. It appeared that the technology used early in the industrialization process was capital intensive, borrowed from the advanced countries. Many thought that a part of the answer to the

Adapted with permission of John Wiley & Sons, Inc. from "Economic Man and Engineering Man: Choice of Technology in a Low-Wage Country," *Public Policy,* Vol. 21 (Summer 1973), p. 313. The field research on which this paper is based was done while the author was a consultant to the Harvard Development Advisory Service. Sjahruddin and Sri Hartono were of great help in the collection of the underlying data. The author is grateful for the comments of Robert B. Stobaugh and Raymond Vernon on an earlier draft.

unemployment problem was to turn to "intermediate technology," technology that involves less capital and more labor than that in general use in more advanced countries.

As pointed out in the previous chapter, if the business manager behaves like an "economic man," he should, according to traditional wisdom, choose either an intermediate or a labor-intensive technology when operating in a less developed country with low wage rates and high capital costs. By choosing appropriately, the manager will minimize the costs of production and simultaneously raise employment.

Most of the literature on the choice of technology fits into one of four categories. Some work deals with the appropriate criteria for a country faced with a choice of alternative technologies.[1] Some discusses the feasibility and costs of alternatives, on the basis of studies of various possible techniques.[2] Other studies ask what can be learned from an examination of aggregated data about the choices made by entrepreneurs.[3] With the exception of the work of Yeoman (reported in the previous chapter), Strassmann, and Mason, before 1972 almost no empirical work was done to examine how the business manager actually behaves.[4] This study of technology in Indonesia has been an attempt to help close the gap.

The principal finding of the research is that the business manager probably does not behave solely as an "economic man." Economics puts a constraint on the manager's behavior, but the manager may also have a competing objective, which causes him to function as what could be described as an "engineering man." For various reasons, managers do not take a narrow profit-maximizing view, and government policies to encourage the use of intermediate technology must take into account this rival function, which appears to influence the design of plants in less developed countries.

THE RANGE OF TECHNOLOGY

The wide range of technologies that exist side by side in Indonesia immediately raises some question about the appropriateness of the model of the economic man. Some of the differences in technology can be explained by the fact that different investors face different factor costs, but there remains a large, unexplained residual.

For this study, data were collected on the technology used in forty-three industrial plants in Indonesia (see Appendix A), including factories in Djakarta, Bogor, Surabaya, Malang, and Medan. In six industries—plastic sandals, cigarettes, soft-drink bottling, bicycle and betjak tires,[5] flashlight batteries, and woven bags—the data covered four or more plants producing comparable products.[6] For this core sample of plants, systematic comparisons of technology could be made. The remaining plants also appeared to fit the patterns observed in the core sample and reported in the tables. The quotations are drawn freely from the larger sample.

As indicated in table 3–1, four of the six industries seemed to include technologies in each of three categories: capital intensive, intermediate, and labor intensive. (Appendix B provides information on how the plants were classified.)

The difference in employment levels under these alternative technologies is rather striking. A good example is the cigarette industry, where the number of workers required to produce one million cigarettes per month is a common measure of automation. In the capital-intensive class, the median of the range of number of workers required was three; in the intermediate class, six; in the labor-intensive class, forty. For the volume of cigarettes manufactured by the firms interviewed, this would translate into 3,009 workers if all firms were capital-intensive, 6,018 if all were intermediate, and 40,120 if all were labor-intensive. Table 3–2, using these crude estimates for the cigarette industry, shows that the choice of technology has similarly important implications for employment in other industries.

TABLE 3–1

NUMBER OF PLANTS, BY TECHNOLOGY AND INDUSTRY

	INDUSTRY						
TECHNOLOGY	Plastic Sandals	Cigarettes	Soft-Drink Bottling	Tires	Flashlight Batteries	Woven Bags	Total
Capital-intensive	2	3	1	1	2	2	11
Intermediate	6	5	3	4	2	2	22
Labor-intensive	0	3	2	1	4	0	10
	8	11	6	6	8	4	43

TABLE 3–2

NUMBER OF WORKERS REQUIRED IN VARIOUS INDUSTRIES FOR VARIOUS TECHNOLOGIES

	INDUSTRY					
TECHNOLOGY	Cigarettes	Flashlight Batteries	Soft-Drink Bottling	Tires	Woven Bags	Plastic Sandals
Capital-intensive	3,009	464	279	—	—	230
Intermediate	6,018	1,392	698	1,207	—	920
Labor-intensive	40,120	5,568	6,417	7,100	—	—

NOTE: Information not available for woven bags.

Different Factor Costs

The observed range of technology in Indonesia could of course be explained by market imperfections that cause different firms to have different factor costs. Each manager may be behaving as an economic man, differing from some other producers in his choice of technology only because of the different costs of labor and capital. The overall data lend some credence to this thesis. However, a more detailed examination sheds considerable doubt on the ability of different factor costs to explain the range of technologies adequately.

True, foreign firms, which are likely to face lower capital costs than do domestic firms, are disproportionately represented among the firms using the more capital-intensive technology (see table 3–3).[7] The two foreign firms listed as using intermediate technology were interesting cases. One had plans to buy new equipment that would put it in the capital-intensive class. Both were using second-hand equipment from their factories in other countries.

TABLE 3–3

NUMBER OF PLANTS, BY TECHNOLOGY AND NATIONALITY OF INVESTOR[a]

	NATIONALITY OF INVESTOR	
TECHNOLOGY	Foreign	Domestic
Capital-intensive	8	3
Intermediate	2	20
Labor-intensive	0	10

[a]If any of the equity was held by a foreigner, the plant was supposed to be registered with the Foreign Investment Board. This registration was the basis of classification.

Three additional pieces of evidence support the idea that the breadth of technology might be a response of an economic man to different labor costs. First, the foreign firms rather consistently pay higher wages than do the domestic firms (see table 3–4). The differences are large for both the skilled and the unskilled jobs and cannot be accounted for by differences in industry mix, location, or payment in kind. Thus, the foreign plant's choice of capital-intensive technology could result not only from access to cheaper capital but also from the fact that the foreigner pays higher wages than the domestic enterprise.

The second bit of evidence comes from the subsidized "medium-term investment-credit program." Those firms that had received money at 12 percent annual interest under this program seem generally to have chosen a more sophisticated technology than their domestic competitors, who are forced to pay the market rate of 24 to 36 percent. One of the three domestic firms using capital-intensive technology was financed under this program, while only three of the twenty domestic firms using intermediate technology and none of the firms using labor-intensive processes had this subsidized capital. Additionally, the state enterprises seem to be consistently more automated than their private counterparts. One of the domestic enterprises employing a capital-intensive process and not financed under the "medium-term invest-ment-credit program" was state owned. Two of the plants employing intermediate technology were state enterprises, while none of those with labor-intensive technology was owned by the government.

The state enterprises may receive subsidized capital; table 3–4 indicates that they face higher wage bills than their private competitors. Their choice of more sophisticated technology may be justified under a purely economic objective function.

TABLE 3–4

Wages in Rupiahs[a] per Day, by Skill Level and Nationality of Investor

	NATIONALITY OF INVESTOR		
SKILL LEVEL[b]	Foreign	Domestic (State)	Domestic (Private)
Unskilled	314	227	123
Skilled	503	425	352

[a]One U.S. dollar equaled approximately 413 rupiahs at the time.
[b]Comparable jobs were taken in each industry. Industry by industry, the same patterns of wage differences were evident.

However, some of the statements of the managers of various enterprises raise questions about the usefulness of the simple model of the "economic man" for fully explaining the decisions that are being made.

The following comments are characteristic:

1. After noting that machines in the process of being scrapped were capable of producing a high-quality product, one foreign manager explained the replacement of the inter-mediate equipment with sophisticated equipment as follows: "You have to modernize to stay ahead of the competition." This statement was made in spite of the fact that there was a ban on the importation of the product and that there were no other firms manufacturing the product locally. When asked what he was going to do with the old equipment, he explained that he was going to cut it up and scrap it. He would not sell it because "some of these Indonesians can get any old equip-ment running."

2. The manager of a hand-rolled–cigarette factory answered the question as to which technology was cheaper by assuring me that automated cigarette factories had the advantage. When asked why he did not introduce machines, he explained that "the interest payments on the money we would have to borrow would exceed our wage bill." When challenged with the possibility that this statement might be inconsistent with his claims on costs, he responded that the automated plant would be cheaper in "the long run."

3. Another manufacturer who was employing young girls to attach labels by hand to his products explained that he was ordering a machine to replace them. Asked whether it was cheaper to attach them by machine, he explained that he did not know, but that the girls were a lot of trouble: "They just cause management problems."

4. The manager of a plant with both an automatic and a semiautomatic line explained that he was converting the semiautomatic to fully automatic as soon as possible. He wanted to produce a "high-quality product." The output of both lines was already meeting the standards of the foreign licensor and was considered by the firm's competitors as

among the best in Indonesia. It was not clear that further automation would improve the quality, at least in a way that consumers would notice.

If factor costs are the principal determinants of technology choices, the engineering data should support the choices by indicating that the capital-intensive technology is suitable for the foreign firm and for the domestic firms with subsidized capital. The data point, however, to a rather different conclusion. At any reasonable set of capital costs that a firm in Indonesia might face, the choice of capital-intensive technology requires an investment per worker not employed that is far beyond that which would be consistent with the actual wages paid by the domestic firm, the foreigner, or the state enterprise.

The Investment to Replace a Worker

One way of determining whether managers are behaving as classic economic men is to calculate the incremental investment made per worker saved (not employed) in one level of technology as compared to the next lower level. The data collected from firms made it possible to perform such calculations for several of the industries.

The model of the economic man appears to be of some value in explaining the choice between the labor-intensive processes and the intermediate level. Unfortunately, only the cigarette and soft-drink bottling industries could be analyzed for this comparison.[8] The incremental investment for machine rolling of cigarettes, as compared to hand rolling, was between $50 and $120 per worker saved. For soft-drink bottling, the additional investment for semiautomatic bottling as compared to hand bottling was on the order of $833 per worker saved.[9]

The wages paid out to cigarette rollers run on the order of $100 per year. Thus, even with the higher investment figure of $120 per worker, a capital cost of 36 percent, and a maintenance or depreciation cost of 10 percent per year, the annual capital cost of eliminating a worker is only about $55. These figures suggest that the hand rolling of cigarettes continues only because of the unavailability of capital, even at 36 percent interest, because of a differentiated excise tax that favors hand-rolled cigarettes, or because the consumer prefers the "low-quality," hand-rolled version.

On the other hand, semiautomatic bottling may well be a poor investment for firms that must pay 36 percent for their capital. The annual capital and maintenance costs would be about $383, considerably above the wage rates in Indonesia. Note, though, that a foreign firm, with an 8 percent interest rate and the same 10 percent maintenance cost, would face an annual capital cost of only $150, which is less than the annual cost of an unskilled worker in a foreign-owned plant. It is thus perhaps not surprising that hand bottling was done by domestic producers only. Semiautomatic equipment was operated by both foreign and domestic enterprises.

In the case of the choice between the intermediate technology and the capital-intensive level, the findings are rather different. Calculations could be made for critical processes in five of the six industries. The average investment per worker eliminated was $6,986; the range was from $2,000 to $21,500 (see Appendix C).

If the foreigner is able to raise capital abroad for 8 percent and the replacement and maintenance requirements amount to 10 percent per year (or a ten-year life for the equipment), the annual wage that would have to be saved by an economic man would, for the average $6,986 investment, have to exceed $1,260 per year. The actual annual cash wage for employees of foreign firms averaged about $225. Even with a high cost for social expenses, it is unlikely that the real costs of a worker could approach $1,260 in Indonesia, nor is it likely that wages will reach that level in the foreseeable future.

At the low end, for an investment of $2,000 per worker, the 18 percent capital costs amount to $360 per year. It is possible that the wages plus social costs of an employee for the foreign-owned plant could reach this level, but in only two industries was the investment below $3,500.

Although the capital costs involved in the capital-intensive technology generally far exceeded any possible wage savings, such investments were frequent and were being contemplated by many firms then using less automated techniques. The simple calculations suggest that these decisions did not make sense from the point of view of an economic man who wants to minimize costs.

Other Economic Factors

Three more complicated economic justifications should be disposed of before the model of the economic man is modified substantially or pushed to the back seat:[10]

1. For the foreign firm, the costs of developing more labor-intensive techniques may exceed the benefits of using known technology.
2. The quality of output may not be sufficiently high or sufficiently consistent with that of the intermediate technology.
3. The intermediate technology may waste raw materials to such an extent that their costs outweigh net savings on capital and labor.

These themes appear so frequently in the literature on choice of technology that they almost certainly have validity in many instances. They did not, however, seem to go very far in explaining what was observed in the light manufacturing plants included in this study.

Where the intermediate technology has to be developed from scratch, or even copied from others, the development costs may well outweigh the benefits to be derived from lower operating costs. One of the plants visited during this study had attempted to introduce a new technology, but, after large expenditures on engineering and construction, the manager found himself with machines that did not function. Moreover, unexpected expenditures on modifications were already under way. In general, however, the managers of the plants included in this study of light manufacturing were well aware of and understood the intermediate alternative. In fact, many of the foreign firms had taken over a going enterprise as a joint venture, or had been given back a plant that had been nationalized under the Sukarno regime. In most such cases, the plant was already operating with an intermediate technology. However, in spite of the economics and their knowledge of alternatives, the firms had already converted or were in the process of converting their plants to a more capital-intensive technology.[11] Ignorance or the cost of developing intermediate technologies was not the problem.

The quality issue is a difficult one; engineering standards and consumer standards of quality do not always seem to agree. Quality was frequently mentioned as a reason for increased automation. However, several facts suggest that the supposed quality differences were either not real or else were not perceived by the consumer as being real.

In one industry, three firms provided engineering test data covering the products of several competitors. The products that ranked highest in level of quality and lowest in variance from this level came from manufacturers whose processes extended well into the range of intermediate technology. One of these plants—one represented in the previous quotes—was operating an automated line next to an intermediate technology line. The output of both was meeting the high standards of the foreign licensor. In other industries, firms were asked to rate the quality of competing products. Again, the highest rankings were for firms that extended well into the range of intermediate technology.

In the case of cigarettes, the engineering standard of quality seemed to conflict directly with the standards of a sector of the purchasers. Machine-rolled kretek cigarettes[12] were having a difficult time taking a share of the market from the hand-rolled ones. The even-burning, crackle-free characteristics of the machine-rolled cigarette seemed to be less liked by the kretek smoker than the uneven, noisy, hand-rolled variety. Also, machines had not been developed to reproduce the slightly irregular, flared shape of the hand-rolled variety, which seemed to be more popular with the consumer.

The labor-intensive techniques did generally result in a product that was inferior, at least by engineering standards, to that of the more capital-intensive technologies. The hand-bottled soft drinks were, for example, probably not terribly hygienic: foreign matter could easily be introduced into the bottles by accident, and flies swarmed around the ingredients before they were hand mixed. The flashlight batteries made through labor-intensive techniques were more subject to leakage than were the others. The case of consumer preference for hand-rolled cigarettes is probably an anomaly. However, the intermediate technologies did seem capable of consistently matching the product quality of capital-intensive techniques.

Contrary to expectations, the saving of raw material was never mentioned by the firms interviewed as a reason for choosing automated production processes; in fact, it appeared occasionally as a reason for not automating. The automatic looms in the bag-making plants produced more rejects than did the semiautomatic ones. Simple manual cutters got more sandal soles out of a piece of foam plastic than did the automatic, ganged ones. And so on. In process industries, saving raw materials may be a justification for automation, but this was not the case with the firms in this sample.[13]

Even these somewhat more sophisticated versions of the model of the economic man seem insufficient to account for the choice of technology. One must look for other explanations.

OLIGOPOLY AND CAPITAL INTENSITY

One impression that was consistently reinforced in the interviews was that the Indonesian entrepreneur wanted as much as his foreign counterpart to have an automated plant. The suggestion was that somehow the foreign firm was in a position to fulfill this desire in more cases than was the domestic investor.

Probably, however, the ability to choose capital-intensive techniques was not so much a function of nationality as of the degree of monopolistic advantage that the firm was able to exercise. Several observations suggest that the firms employing capital-intensive technology were more likely to enjoy monopoly positions. First, all but one of the foreign firms interviewed were using international trade names. The one exception had been operating in Indonesia for more than thirty years; it had developed brands that it did not use in other countries but that were *thought* by most consumers to be international. These brand names had put the foreign firms in a position where price was not the basis of competition. On the domestic side, monopoly also played a role. Of the three domestic firms using advanced technology, two had monopolistic positions. One was using an international trade name under license; the other was a state enterprise that sold almost all of its output to another state enterprise.[14] In the intermediate category there were only two domestic firms using international trade names. Both of the plants were noticeably more automated than the other domestic firms in the same industry.

Price serves as another indicator of monopoly position. The average price of the product of the capital-intensive plants was far higher than that of the intermediate-level product in all cases for which data were available (see table 3–5). Although the firms with capital-intensive technology all produce high-quality products, these products do not appear to rank higher in level of quality or consistency than those of the best of their competitors without international trade names. Their prices, though, are higher.

Further hints that monopolistic advantage—not foreign status—underlies capital intensity are given by the cases of foreign firms

TABLE 3–5

APPROXIMATE AVERAGE RETAIL PRICE OF PRODUCT, IN RUPIAHS

TECHNOLOGY	INDUSTRY				
	Plastic Sandals[a]	Cigarettes[b]	Soft Drinks[c]	Bicycle Tires[d]	Flashlight Batteries[e]
Capital-intensive	120	78	63	470	55
Intermediate	60	60	36	350	40

[a]One-piece sandals, per pair.
[b]The most popular brand, for each plant, per package of twenty, white cigarettes only.
[c]For the contents of a bottle, adjusted to reflect volume.
[d]Wholesale.
[e]Size "D" (UM-1).

outside the industries studied systematically.[15] In these cases, it seemed that where trade names were not important, firms appeared to employ a much more labor-intensive technology than they used at home.

THE DRIVE TOWARD AUTOMATION

Several factors seem to drive firms toward more capital-intensive techniques than would be suggested by the simple economic models. In many cases the managers appear to want to respond to certain elements of risk and uncertainty. Capital-intensive plants seem, in some instances, to serve an insurance role. And some managers seem to be driven not only by simple economic objectives but also by objectives that could perhaps be best described as those of an "engineering man." But the ability of a manager to pay the cost of insurance against risk and uncertainty, or to let the engineering objectives have their play, is much greater if the firm has a monopolistic position than if it faces severe price competition.

There are at least two kinds of insurance coverage that a capital-intensive plant might offer that the labor-intensive plant does not. The capital-intensive plant may allow the manager to respond more quickly to unexpected fluctuations in demand or to levels that are different from that for which the plant was designed; and the capital-intensive plant may be perceived as allowing the manager to reduce the risks of future liquidity crises.

Consider the problem facing an entrepreneur in whose market the level of sales is uncertain or is likely to vary considerably. If the technology that minimizes costs is one that makes it difficult to adjust plant output to levels other than the optimal design output, the entrepreneur may decide not to adopt it; he may instead choose a technology that allows for easy adjustment of output, even though it may entail a higher unit production cost.

In many cases, the more capital-intensive techniques are more readily adjusted to different levels of demand. Confronted with sporadic and unpredicted increases in demand, the manager can more easily run an added shift in an automated plant than he can recruit and train the necessary complement of workers in a labor-intensive plant. With brief contractions of demand, automated plants can be adjusted by running the machines more slowly. In labor-intensive plants, the alternatives of laying off workers or slowing the work pace are particularly unattractive. Lay-offs are expensive or difficult in a country such as Indonesia, where the labor laws discourage the practice. And slow-downs may result in work habits that are hard to change when production is to be expanded again. Of course small fluctuations in demand may be handled out of inventory manufactured at constant output levels. However, in many cases the storage costs and the risks of product obsolescence or deterioration will severely limit the manager's ability to use inventory to match a constant output to a fluctuating demand.

If the insurance against errors in demand projections is attractive, perhaps the firms facing price competition should behave as the monopolists do. However, an entrepreneur faced with keen price competition must be concerned that his competitors may decide not to pay insurance premiums—that is, the higher production costs of capital-intensive technology. In that case, the insured firm faces a lopsided loss-gain situation. If the original market forecasts turn out to be correct, the insured firm may be driven out of business by those not having to bear the insurance costs. At worst, the uninsured firm may miss some opportunities to profit from a quick expansion of its ouput. The entrepreneur's response, therefore, may be not to pay the insurance but to train a work force capable of producing at a level on the low side of his projected sales.

On the other hand, the firm with a monopoly rent that derives from its inelastic demand curve may be able to afford the insurance. It can bear the higher production costs at the expected demand level and expand or contract easily if the market turns out to be larger or smaller than that for which the plant was designed to be optimal.

When the monopolistic firm is vertically integrated, the gains in being able to respond to an unanticipated increase in demand may be larger than is indicated by the profits of the operation. If the operation is only an assembly, bottling, or packaging plant, the increase in demand opens up the possibility of selling added inputs from existing capacity in other plants. The output of such plants will, as a rule, decline in unit cost as volume increases. The design of a foreign-owned pharmaceutical plant provides a good example. The plant makes tablets and capsules from the products of factories located elsewhere but owned by the same parent. The output of this capital-intensive plant can be expanded easily to meet unanticipated demand, and the incremental contribution to the multinational system of plants from the sale of a bit more chemicals is very high. The higher costs of the capital-intensive plant at the final stage can easily be offset by the additional profits that derive from the use of more intermediates from affiliated firms whenever demand rises.

The second kind of insurance that a more capital-intensive plant may offer is that the chances of a liquidity crisis will be reduced. Managers responding to this kind of issue claimed that capital-intensive plants were "cheaper in the long run," in spite of the fact that an economist would argue that the labor-intensive operation had lower costs. The managers' behavior is at least partly a result of the imperfections in capital markets.

Consider the firm that now enjoys monopolistic profits but fears that price competition may become so severe in the future that prices will fall close to marginal costs. Prices may even approach the marginal costs of a capital-intensive manufacturer rather than the higher marginal costs of a labor-intensive firm. The entrepreneur in Indonesia is constantly reminded of this possibility by some plants that had to close as a result of intense price competition.

Faced with this possibility, the businessman might decide to build a labor-intensive plant with minimal costs and invest the incremental cash in assets that would yield a return. If prices fell below his marginal costs later, he could close the plant and still have an income from the assets obtained thanks to his lower costs during the period of high prices. This income would presumably be as great as the cash flow that the competition with a capital-intensive plant would derive from staying in business. But two factors make this strategy unattractive. First, the businessman perceives a high risk attached to alternative investment for his assets in imperfect capital markets; moreover, he is reluctant to withdraw from his basic business or to remain in a business with a cash deficit that would have to be subsidized.

An alternative strategy would be to borrow to build a more capital-intensive plant at the beginning, pay off the loan during the period of high profits, and stay in the same business during the competitive period as long as prices remain above his low marginal costs. This strategy generally appears much more attractive than that of minimizing costs. If intense price competition becomes a reality, the entrepreneur can stay in the same business without having to make up a cash deficit, and he can avoid the risk of having to invest assets in the capital market to cover later losses.

Although the strategy that minimizes cash costs in the future appeared to be more attractive to most Indonesian businessmen, the only ones who could follow it generally were those with a sufficiently monopolistic position to allow them to bear the burden of financing the capital-intensive plant.

While the drive toward capital-intensive technology may be partly explained by the desire of the oligopolist to insure against risk and uncertainty, it seems also to be a response to some objectives of the engineering man. These are:

1. Reducing operational problems to those of managing ma- chines rather than people.
2. Producing the highest-quality product possible.
3. Using sophisticated machinery that is attractive to the en- gineer's "aesthetic."

Of course, reducing management problems can be viewed as an economic objective if management is considered a production factor, and if management attention freed from personnel problems could be devoted to other money-making activities, with a higher return. However, there is no reason to expect the opportunity cost of managers of plants in an oligopolistic position to be higher than those of plants in a more competitive market. The drive appears to be only partly a response to economic opportunity costs.

Similarly, the production of a high-quality product could be viewed as an economic objective if quality increased sales sufficiently to offset the higher production costs. But the managers' concept of quality appeared to go far beyond that which had a meaning to the customer. And, although variances in quality could be a major problem in the very labor-intensive technologies, it did not seem to be an issue with the intermediate techniques.

It should not come as a surprise that engineering objectives play a role in the choice of technology. After all, most plants are designed by engineers. Nor is it surprising that engineering objectives should sometimes depart from economic objectives; it has long been recognized that scientists respond to objectives not easily captured in economic models. Engineers also have a training that may inculcate noneconomic values, and responsiveness to these values need be no less rational than is a response to economic values.

It is worth noting that few of the Indonesian firms did any calculations to investigate the technology appropriate to them. On the other hand, most of the foreign firms required some kind of feasibility calculation before new equipment was installed, but such studies did not generally compare one technology to an alternative. The most common type of calculation was an engineering study in which the economic return was calculated only for the equipment proposed by the engineers; no comparison with alternative techniques was made. In fact, in one case the calculations were done in the United States by a feasibility team composed of engineers who had never visited Indonesia. The suggestion was that U.S. wage rates might have been used in the calculations. The manager explained, when asked about the effect of lower wage rates, that "you get what you pay for in productivity."

CONSTRAINTS ON AUTOMATION

If a firm is forced to compete on a price basis, economics constrains the manager's ability to allow engineering objectives to override economic ones, as well as the firm's ability to use capital-intensive technology to insure against risk and uncertainty.[16] Our research indicated that there are other constraints as well. In fact, only one of the sophisticated plants was described by management as being as automated as would be a similar plant in an advanced country.

The principal constraint for the firms with inelastic demand seemed to be the problem of maintaining and operating advanced equipment in a developing country. In fact, one domestic firm had installed very sophisticated machinery but was gradually disconnecting the automatic controls as they broke down. And most of the complaints about the quality of Indonesian labor came from the plants with the most capital-intensive technology; maintenance and operation of the sophisticated equipment were requiring scarce skills.

The labor laws appeared to provide some brake for firms already in operation. A few managers explained that they had definite plans to automate, but could not do so immediately because they were not allowed by the government to release the workers who would thereby become redundant. As soon as natural attrition of workers enabled them to do so, they would proceed to a more sophisticated technology.

The inability to raise capital at any price seemed to prevent some of the very labor-intensive firms from moving to intermediate technology. Some were unable to borrow from the state banks and did not have the contacts to raise money on the quasi-legal private money market.

CONCLUSIONS

This study has only begun to probe the complex factors that influence managers in their choice of technologies. As the previous chapter concluded, the simple combination of production functions and factor costs is clearly inadequate to the task. But there are many promising hypotheses waiting to be tested. A surprisingly wide range of technology was found to exist within the same industry and country. The breadth is probably partly a result of the fact that the study examined only light manufacturing. However, the factors that influence the managers in their choice of technology are probably similar to those in other kinds of industries.

The capital-intensive techniques were generally associated with foreign investors and, more significantly, with investors who had some monopolistic position. In the terms used in the previous chapter, the firms faced relatively price-inelastic markets. Foreigners happened to be in such a position more frequently than domestic investors. However, the drive toward capital-intensive plants seemed to be similar among both foreign and domestic managers.

The managers' choice of technology appears to be influenced by two objective functions, which, in low-wage countries, generally conflict. The first objective, that of the "economic man," is to minimize costs, which leads to a relatively labor-intensive production process. On the other hand, the objective of the "engineering man" tends to lead toward more sophisticated, automated technology. Where price competition is the rule, the objectives of the economic man seem to override those of the engineering man. But when the firm has a monopolistic advantage, it is under less pressure to minimize costs to survive; the goals of the engineering man can be allowed to lead the firm to a level of technology more advanced than that which the economic man would choose.

Additionally, capital-intensive plants appear to provide some insurance against risk and uncertainty. Under some circumstances, only the firms that have a monopolistic position can afford to pay the insurance premium—in the form of higher production costs—associated with the capital-intensive choice.

The relative importance of the insurance factors and the engineering objectives has not yet been examined.

In more advanced countries there may be little conflict between the economic and the engineering objectives. With high labor costs, automation is the direction sought under both objectives, and insurance is gained automatically.

Macrostudies have accumulated considerable evidence that the average capital-to-labor ratio for individual industries in the high-wage countries is greater than that in the low-wage countries. It is also possible that low-wage countries employ a wider range of technology.

Although relative factor costs play a role in the selection of technology, there appear to be other influences on the managers' decisions that are of critical importance to the policy maker interested in the employment effects of industrialization.

APPENDIX A

AGE AND SCALE OF PLANTS

Two of the difficulties frequently encountered in cross-sectional comparisons of technology were minimal in this study: the inclusion of old plants along with new ones, and the problem of economies of scale.

Of the forty-three plants studied, twenty went into production in 1965 or later. Of the twenty-three plants that were established earlier, nine had made substantial additions to their equipment since 1965. The more recently introduced technology was used as the basis of classification. Six of the remaining old plants were in the cigarette industry. Although the foreign plants accounted for the same percentage of old plants as they did of the whole core sample, none was represented in the plants that had no major additions since 1965.

There was considerable overlap in the size of plants in the capital-intensive and intermediate categories. In all but one of the industries, some plants using intermediate technology had an output at least as great as one of the plants using capital-intensive technology. And in three of the industries, the capital-intensive plants and the intermediate plants had several machines at the principal stage running side by side. However, the labor-intensive plants were generally considerably smaller than their counterparts with a more advanced technology.

APPENDIX B

METHOD OF CLASSIFICATION

Technology can be classified in various ways. Most popular, in studies at the macro level, is the use of capital/labor ratios. In this study an attempt to use capital/labor ratios at the firm level was quickly discarded, because of several difficulties. The principal one was the problem of finding an appropriate measure of the value of the capital employed: the bookkeeping of the firms interviewed was not consistent; different rates of depreciation were used; purchase at different times led to different values for undepreciated equipment; book value sometimes seemed to reflect tax and duty considerations, especially when equipment was purchased through a foreign affiliate; and books kept in rupiahs were useless where the life of the firm extended back into the period of rapid inflation in Indonesia (moreover, firms were hesitant to show their books). The second problem came from the fact that the degree of vertical integration was not the same from firm to firm. Although the number of workers could be adjusted, approximately, to reflect a common level of integration, the capital figures were terribly difficult to disaggregate. An additional difficulty lay in accounting for the "inferior technology" found in a few cases. In these instances, both the capital and the labor required to produce a given output were greater than that under some alternative technology. Although such technology was not encountered in many cases, a mechanistic measure might have distorted the findings.

As a result of these difficulties, plants were classified principally by the kind of equipment in use. (Where new equipment was actually on order, it was counted as being in use.) Some key processes were identified in each industry that seemed to be good reflections of the level of automation throughout the plant. These classifications were checked, to the extent possible, by using labor/output ratios, where labor was adjusted to comparable levels of vertical integration. The labor/output ratios were crude, but they permitted some approximations of the employment implications of the various levels of technology.

The categories of technology (capital intensive, intermediate, and labor intensive) obviously are rather arbitrary. To protect the confidentiality assured the firms, some way had to be found to aggregate the industries. The author hopes that the reader will take it on faith that the disaggregated data contained no surprises.

The classifications were based on the following criteria:

Cigarettes

- *Capital intensive:* Use 2,000-per-minute cigarette makers, and machines for all stages of tobacco preparation. Average of three workers per million cigarettes per month.
- *Intermediate:* Use of 1,300-per-minute or slower cigarette makers. With one exception, some hand preparation of tobacco. Average of six workers per million cigarettes per month.
- *Labor intensive:* Hand rolling of cigarettes and primarily hand preparation of tobacco. Average of forty workers per million cigarettes per month.

Flashlight Batteries[17]

- *Capital intensive:* Linked conveyor system, most of the parts assembled by machine. Average of one worker per thousand batteries per day.
- *Intermediate:* Unlinked conveyor system, some parts assembled by machine. Average of one worker per thousand batteries per day.

- *Labor intensive:* Unlinked conveyor system or no conveyors, all assembly by hand. Average of twelve workers per thousand batteries per day.

Soft-Drink Bottling

- *Capital intensive:* Automatic uncrating of bottles, loading of bottle washer, and crating of bottles. Average of one worker per thousand bottles per day.
- *Intermediate:* Bottling machine, but crates unloaded and loaded by hand. Average of 2.5 workers per thousand bottles per day. (There was a considerable difference between the top and bottom ends of this category.)
- *Labor intensive:* Hand washing and hand filling of bottles. Average of twenty-three workers per thousand bottles per day.

Tires

- *Capital intensive:* Use of Banbury mixer, automatically timed curing molds, and completely power-driven calenders. Automatic bead-making and bias-cutting equipment. Unable to estimate average number of workers.
- *Intermediate:* Use of millers, partially hand-driven calenders. Un-timed curing molds. Semiautomatic or hand mixing of beads and bias cutting. Average of 8.5 workers per hundred tires per day.
- *Labor intensive:* Most operations by hand. Average of fifty workers per hundred tires per day.

Woven Bags

- *Capital intensive:* Use of 200-pick-per-minute automatic looms, one operator per four to six looms.
- *Intermediate:* Use of 150-pick-per-minute or slower looms, one operator per loom. Overall workers-per-unit output could not be calculated because of the use of different fibers, requiring different preparation.

Plastic Sandals

- *Capital intensive:* Use of automatic injectors, requiring one to 1.5 operators per injector. Approximately five workers per thousand sandals per day for simple sandals.

- *Intermediate:* Semiautomatic injectors, with two to five operators each. Approximately twenty workers per thousand sandals per day for simple sandals.

APPENDIX C

INVESTMENT PER WORKER

To determine the investment per worker saved by employing the intermediate technology instead of the labor-intensive technology, the processes described below were compared. In each case, the cost data were collected from firms that were using the technology, and the numbers of machines and workers were adjusted to reflect a given output level.

1. *Cigarettes:* Use of second-hand, 1,300-per-minute makers or 2,000-per-minute makers, modified to handle clove mix, in comparison to hand rolling with a simple wooden roller. Second-hand machines appeared to be readily available, from abroad or from local plants that were automating.
2. *Flashlight batteries:* Insufficient data available.
3. *Tires:* Insufficient data available.
4. *Soft-drink bottling:* Use of a simple semiautomatic bottling machine in comparison to hand filling of bottles.

To determine the incremental investment per worker saved by employing the capital-intensive technology instead of the intermediate, the following processes were compared:

1. *Cigarettes:* The use of new, 2,000-per-minute makers in comparison to the use of second-hand, 1,300-per-minute machines.
2. *Plastic sandals:* The use of automatic injectors in comparison to the use of new semiautomatic injectors from Hong Kong.
3. *Flashlight batteries:* The use of an automatic conveyor assembly system from Japan in comparison to the use of a semiautomatic system from Japan.
4. *Woven bags:* The use of fully automatic looms in comparison to the use of semiautomatic flat looms.
5. *Soft-drink bottling:* The use of automatic crating equipment in comparison with the use of manual crating.
6. *Tires:* Insufficient data available.

4

More on Production Techniques in Indonesia

JAMES KEDDIE

$\mathbf{I}$n the previous chapter, Wells argued that the conventional economic model could not adequately explain the production technology choices actually made in developing countries. True, he acknowledged that at times firms choose capital-intensive technology for economic, albeit difficult to measure, reasons: to reduce the risks of liquidity problems and any errors in matching production to demand. But his central argument was that the predilection of managers for capital-intensive techniques may be due to the motivations of the "engineering man," which keep managers from be-

This chapter is based on the author's "Adoptions of Production Technique by Industrial Firms in Indonesia," Ph.D. dissertation, Harvard University, 1975.

having purely as "economic men." This chapter, in contrast, reports Indonesian field work that suggests considerable support for the view that choices of production technology have been economically motivated.

Product differentiation can be very important to business firms in developing countries. Wells noted that the products of the capital-intensive firms commanded substantial price premiums—from 30 to 100 percent—over those of the firms using intermediate technology, in each of five industries studied.[1] Although brand names and advertising may have contributed to these differentials, it seems likely that more objective standards of product quality played a part as well. In the two industries in which brand consciousness is probably most important (cigarettes and soft-drink bottling), the average price premium was actually lower than in the remaining three (batteries, tires, and plastic sandals).

Cost differences among the competing technologies are unlikely to have been as large as the differences in price realization. An interesting parallel is suggested by the work of Strassmann on cost differentials among techniques of cotton textile production in Latin America. He found a cost spread per meter of only 9 or 11 percent (depending on the rate of interest assumed) between the different manufacturers studied.[2] The results of Wells and Strassmann, of course, are not directly comparable; different industries and countries are involved. But the evidence suggests that, with marked product differentiation in developing countries, product price premiums may considerably exceed unit cost differentials across techniques of production with which the premiums are associated. Thus the field is open for two hypotheses:

1. The primary concern of firms in adopting techniques is to secure a premium quality or other objective product advantage; and
2. This concern is a response to the economic opportunities and risks generated by product differentiation and heterogeneity, representing an economically motivated pursuit of objective product advantage.

The remainder of this chapter is largely devoted to tests of these hypotheses.

CONCERNS IN ADOPTING TECHNIQUES

The research on which these tests are based was carried out in Indonesia in 1973. The main method of gathering data was personal interviews with executives in private firms, both foreign-owned and Indonesian-owned. Information thus provided was supplemented by interviews with technical experts, advisors, and agents of equipment suppliers in Indonesia and subsequently by correspondence and by some additional interviews when the author revisited Indonesia in early 1975.

The sample of firms was drawn from nine manufacturing and two construction industries. In manufacturing, the industries were bricks, cardboard boxes, cigarettes, flashlight batteries, paint, printing (of packaging materials), soap, textile printing, and weaving. The construction industries were large concrete-frame buildings and roads. For each industry, the selection of firms was designed to cover a wide variety of alternative techniques. Large, prominent firms were interviewed first, smaller firms usually being identified in discussions with the large firms. About one-third of the firms were owned or technically dominated by investors from developed countries; the remainder, by Indonesians or investors from other Southeast Asian countries.

Both qualitative data (reasons for adopting techniques) and quantitative information (operating and financial statistics of firms and techniques) were sought during interviews. Because of the volume and sensitive nature of the quantitative data sought, a formal questionnaire was not used. Instead, the interviewer kept in mind a list of topics throughout each interview, raising each topic as occasion permitted.

To test the hypothesis that firms' primary concern in adopting techniques is to secure an objective product advantage, executives were asked why they had adopted the techniques they used, as opposed to alternative techniques used in Indonesia. (The approximate range of techniques used in each industry—from automatic to highly labor intensive—was determined by preliminary inquiry of a large foreign firm or a technical expert.) In subsequent interviews, simple descriptions of alternative techniques were sufficient to allow executives to identify them and to give definite reasons why they had preferred their own methods.

Adopted techniques were subsequently categorized as high- or low-investment cost. The categorization depended on whether, given

1973 equipment prices and operating output rates, the equipment investment was higher or lower than would have been required, for the same output, if the firm had chosen a particular alternative technique. The author chose the specific techniques used for these comparisons so as to ensure that decisions concerning a number of both "high"- and "low"-investment techniques would be included in the sample.

This criterion was applied to a total of eighty-eight discussed decisions about techniques; almost two-thirds of the outcomes (fifty-seven decisions by thirty-one firms) favored high-investment-cost techniques; thirty-one decisions (by thirty firms) were to adopt low-investment-cost techniques. The two categories of decision were then separately analyzed. It had been hypothesized that a prominent reason for adopting low-investment-cost techniques would be the firms' perceived inability to raise the funds needed for higher-investment-cost alternatives.[3] The pattern of reasons for adoption, therefore, was expected to differ for high- versus low-investment-cost techniques. Firms adopting high-investment-cost techniques presumably were less troubled, if at all, by scarcity of funds. Their reasons for technology choice could therefore be expected to show much more directly and clearly the postulated primary concern for objective product advantage.

Reasons for the adoptions of high-investment-cost techniques were classified under four headings: quality, labor-related problems of maintaining output, machinery-related problems of maintaining output, and economy. There was some overlap between the categories. For example, even though physical quality is the most obvious form of objective product advantage, firms might also perceive that rapid or assured delivery was important to their commercial success; but references to such problems as those of maintaining output from large-scale applications of manual techniques were classified as "labor-related problems of maintaining output"; and difficulties in obtaining spare parts or service from suppliers of equipment for alternative techniques were classified as "machinery-related problems of maintaining output." Finally, comments indicating a concern to cut wage or raw material costs, or even depreciation costs (through purchase of longer-lived equipment), were classified under "economy."

All firms mentioned more than one reason in support of their choice of technology; in fact, there were 124 reasons given for the total of fifty-seven high-investment-cost decisions. Two alternative classification procedures were used: 1) decisions were tabulated under the heading of the first-mentioned reason for each decision; or 2) decisions were tabulated under the heading that covered a plurality (as opposed to a majority) of the reasons mentioned.

Under either classification procedure, concerns for product advantage were cited much more often than economy as reasons for adopting high-investment-cost techniques (see table 4–1). This pattern holds for both manufacturing and construction, although quality—as opposed to maintenance of output—is much more clearly the form of product advantage sought by manufacturers. This is plausible, as manufacturers normally produce to their own chosen specifications; whereas, on construction projects, designs and workmanship tend to be specified and monitored by architects and consulting engineers on behalf of the client. Thus, completion dates and the penalties for not meeting them might well be expected to be more important to contractors than to manufacturers.

TABLE 4–1

REASONS FOR ADOPTIONS OF HIGH-INVESTMENT-COST TECHNIQUES, 57 INVESTMENT DECISIONS[a] BY 31 FIRMS IN INDONESIA (PERCENTAGE OF TOTAL DECISIONS)

REASON FOR ADOPTION	MANUFACTURING		CONSTRUCTION	
	First Mention (46 Decisions)	Plurality of Mentions (43 Decisions)	First Mention (9 Decisions)	Plurality of Mentions (7 Decisions)
Quality	57%	60%	44%	43%
Labor-related problems of maintaining output	9	7	33	29
Machinery-related problems of maintaining output	7	7	0	0
Economy	28	26	22	29
Total[b]	100%	100%	100%	100%

[a]There were a total of 57 decisions, each of which was classified by the two alternative methods shown in the table. Two decisions were eliminated from the "first-mention" classification because "first-mention" was split between two headings; seven were eliminated from the classification of "plurality of mentions" because there was no clear plurality under any heading.
[b]Columns do not necessarily total 100 because of rounding.

The dominance of concern for product advantage becomes still greater if each decision is weighted by the incremental investment involved in adopting the high-investment-cost technique as compared to the discussed alternative technique. This procedure, of course, gives greater weight to decisions where more investible funds are at stake. In manufacturing, adoptions for quality reasons alone accounted for two-thirds to three-quarters of incremental investment (depending on the classification procedure); one of the two "maintaining output" classifications accounted for a further 5 to 10 percent. In construction, more than 90 percent of incremental investment was accounted for by a single decision: the adoption of heavy earthmovers for a large Sumatran road project. This decision was justified as a means of avoiding "labor-related problems of maintaining output" with manual techniques. Thus, product advantage appears far more important than economy as a motivation for incremental investment in high-investment-cost techniques.

For low-investment-cost techniques, the picture is more complicated. Because it seemed likely that a shortage of funds was often a crucial factor in these decisions, a two-stage procedure of classification was adopted. First, the decisions were classified into two groups: 1) those explained by the set of categories already noted (suitably extended, e.g., by including under "economy" references to use of cheap labor); and 2) those explained by a shortage of funds. This latter group included two basic kinds of reasons: a) an inability to obtain adequate funds for the discussed alternative technique at any price, and b) prohibitive financial risks involved in the required incremental investment. As anticipated, "shortage of funds" was commonly cited as a reason for low-investment-cost decisions, accounting for twenty of a total of fifty-four mentions for the thirty-one such choices.

In a second-stage analysis, the shortage-of-funds decisions were reclassified to determine what would have happened if the availability of funds were not an issue. The reasons were reclassified under the product advantage or economy headings when the firms in question specified benefits associated with the discussed alternative techniques. Some three-quarters of the mentions could be reclassified in this way. The benefits specified were usually product advantages rather than economy.

Each decision was then given equal weighting, using both the "first mention" and the "plurality of mentions" procedures already described. In contrast to high-investment-cost decisions, these choices were not primarily motivated by a desire for product advantage. In both the manufacturing and the construction sectors, economy accounted for at least as many decisions as did product advantage. As a rule, however, the economy-concerned firms were not deliberately pursuing low-cost, low-product-advantage strategies. One, for example, was the highest-quality producer in its industry, but was taking advantage, for reasons of economy, of a major capital-saving innovation that had no significant effect on product advantage. Others were small construction firms unable to bid for the larger, more profitable contracts that could provide financial coverage for the purchase of heavy equipment thought to be more economical on large contracts; in such a case, economy and product advantage went together. Still others were weavers, choosing cheap looms for economy at the lower end of their product lines, but not neglecting the higher-quality markets that accounted for the bulk of their revenue and output. As a final example, one firm was a medium-quality soap producer that was threatened by the introduction of detergent powders into Indonesia; it had countered with a highly original product innovation—a detergent cream that could be manufactured without a large investment. It was an immediate commercial success, and the firm had thus been able to upgrade its product line substantially and economically.

Thus, very few of the economy-concerned firms were willingly pursuing strategies of economy and low-quality products. Rather, the normal pattern was to seek, where possible, a product advantage and to choose techniques accordingly. This concern is most clearly and directly shown in the reasons for adopting high-investment-cost techniques, but it also appears strongly—if often in a constrained or modified form—among the adopters of low-investment-cost techniques.

A further test, moreover, suggested that the expressed concerns for product advantage were not mere rationalizations (pride, for example, in sophisticated equipment) but were technically valid. Two questions were asked. First, did the manufacturer's concern for quality actually lead to a quality advantage over firms using the discussed alternative techniques? (The question was limited to manufacturing firms, in which quality is relatively easy to assess.) Second, in the opinion of technical experts, were the techniques chosen for product advantage actually likely to produce such advantage, in terms of quality and maintenance of output, over products produced with the discussed alternative techniques?

Quality rankings in manufacturing were established either on the basis of personal observation, including physical handling of the products, or through manufacturers' comparative product tests, or in on-the-spot consultation with independent technical observers (such as resident U.N. experts). In the large majority of cases (88 percent), firms seeking superior quality did achieve rankings higher than those of firms using the discussed alternative techniques. In only 4 percent of the cases did their rankings fall lower; in the remaining 8 percent, quality rankings were the same for both techniques. Thus, quality advantages were generally achieved by the manufacturers who sought them in choosing their techniques.

To assess the importance of the chosen technology for product quality, an individual or institutional source of expert opinion in the United States or the United Kingdom was consulted. These experts were either identified on the basis of technical literature or were recommended by individuals who were initially contacted but who felt themselves technically unqualified on the issue. In the cigarette and printing industries, for example, experts working for equipment-supplying firms were cited as the most competent to appraise the technical choices posed. (None of the experts was asked to judge a case in which his own firm's equipment was involved.) In the remaining nine manufacturing and construction industries, the sources were research institutes or associations, university or government departments, or independent individual consultants. It was thus possible to keep the judgments generally free of institutional bias in favor of specific techniques, particularly since the experts were not informed of the hypotheses underlying the research, nor were they told which were the adopted and which the discussed alternative techniques.

This inquiry revealed a consistent pattern both across sectors (manufacturing and construction) and across the firms concerned with quality and/or maintaining output. For each of the specified product advantages, a large majority of the experts (from 86 to 100 percent) thought that the technique adopted would have a favorable effect as compared with the discussed alternative technique.

The first hypothesis of this research was thus strongly supported. That is, the primary concern of firms in adopting techniques is product advantage rather than economy. This concern was expressed by the firms interviewed and it appeared to be technically valid.

THE MOTIVATION OF CONCERN FOR PRODUCT ADVANTAGE

The second hypothesis of this research was that the concern for product advantage represents an economically motivated response to the opportunities and risks generated by product heterogeneity and imperfectly substitutable techniques. The test of this second hypothesis involved comparing the net product-associated benefits (NPB) and the net technique-associated costs (NTC) of the advantage-concerned firms.

To calculate NPB and NTC for each decision, the firm's actual economic situation (product revenues and costs of raw materials, advertising, equipment, labor, etc.) was compared with a hypothetical situation in which the same firm produced at the same scale of physical output but used the discussed alternative technique and had the product price levels of the rival firms actually using the alternative technique. NPB was then calculated as the difference between the two situations in sales revenue minus the difference in raw material and advertising costs (since these costs could make an independent contribution to product advantage levels). NTC, on the other hand, was calculated as the difference between the two situations in the costs of buildings, equipment, spares, fuel, and labor—the combined costs associated with the techniques. In the actual situations, the product and input price levels, factor productivities, and plant utilization rates of the advantage-concerned firms were assumed to prevail; in the alternative situations, those of the firms using the discussed alternative techniques were assumed to prevail.

NPB was taken to represent the economic stake in the product market—the benefits associated with higher product prices and revenues—that would be at risk if the discussed alternative technique were substituted for actual current practice. NTC was the cost differential to the firm of using its adopted technique as opposed to the discussed alternative. The greater the difference between NPB and

NTC, the stronger the case for an economic motive in technology choice.

Several factors limited the number of decisions that could be examined in this way. Operating data for some firms (including all those in the construction sector) were insufficient; moreover, some firms used heterogeneous technologies that made it impossible to isolate the NPB associated with the use of adopted versus discussed alternative techniques. Comparisons of NPB and NTC were possible, however, in thirty-four instances in seven manufacturing industries—cigarettes, flashlight batteries, paint, printing, soap, textile printing, and weaving.

NPB and NTC were evaluated by discounted-cash-flow methods over ten years of operations after the adoption of the techniques. Since all adoptions were made between 1967 and 1973, it was necessary to project both prices and output to cover ten-year spans. Price indices for various production inputs were selected or constructed, as appropriate. For example, the indices for equipment from various countries of origin were those countries' official engineering price indices; indices for labor and for cigarette raw materials were averages of indices provided by three major firms in Indonesia from their experience. Regression estimates of rates of inflation of the various prices were then based on the index series, which covered the period from 1968 to 1973.

Product-price projections were then derived from the raw material price indices. It could not be assumed that product prices would necessarily rise in proportion to raw material or other costs, because as Indonesian import substitution intensified and more firms entered each industry, competition might be expected to squeeze margins. Also, as the market grew, scale economies might lower costs. Indeed, a margin squeeze was experienced by the firms considered. Their most important cost element was raw materials, and the ratio of sales revenue to raw material costs (typically about 2.0 to 1.2) had been declining in nearly all cases. Where ratios for a firm were available for a span of four or more years, the trend was estimated by regression analysis and superimposed on projected raw material prices in the industry, to yield product-price projections; where less than four years' ratios were available, trends were estimated on the basis of ratio series in each industry from census records on Singapore. (This

neighboring country had undergone a similar import-substituting industrialization a few years previously.) Series from about the mid-
1960s to 1972 were used, and the range of ratios of revenue to raw
materials was very similar to that found in Indonesia.

Output was generally projected at 1973 levels for the advantage-
concerned firms; 1974 was a year of recession in Indonesia and thus
atypical for projections into the medium-term future. A few firms,
however, were still expanding their operations in 1974 from a recent
start-up. For these firms, 1974 output levels were used as the projected
levels. The levels of projected plant utilization in the actual situations
of the firms were also those prevailing in the base year, 1973 or 1974.
For each alternative situation, the standard rate of utilization was that
prevailing in the firm using the discussed alternative technique in 1973
(1974 in one case with a 1973 start-up).

The discount rates used in the analyses were those appropriate to
each advantage-concerned firm. For multinational firms, they were the
weighted average costs of capital derived from consolidated accounts,
averaging 9.5 percent for six firms, with a range from 2.3 percent to 18.3 percent. Published accounts were not available for the
Indonesian-owned firms, but a weighted average cost of capital could
be derived from data supplied by these firms. The average cost of
capital of the Indonesian firms was 17.4 percent—much higher than
the mean for the multinational firms, largely because the Indonesian
firms required very high rates of return on their own equity funds.

Using these methods, it was possible to evaluate eighteen decisions to adopt techniques and thus to make thirty-four comparisons of
NPB with NTC. The number of comparisons exceeds the number of
decisions because, in many cases, two or more firms were using the
discussed alternative technique, and the operations of each of these
firms—its product price levels, factor productivities, and so on—
formed the basis of a separate alternative situation and thus a separate
comparison with the advantage-concerned firm's actual situation.

A two-stage procedure was used to test the hypothesis of the
economic motivation for decisions. First, each comparison was
classified as to the degree of its support, in isolation, for the hypothesis; then the classifications of the comparisons pertaining to each
decision were used to determine the decision's level of support for the
hypothesis. The range of types of relation between NPB and NTC was

quite large; either or both of these measures could be positive or negative, though in fact NPB was positive in an overwhelming majority (thirty-two) of the cases and NTC was positive in twenty-three of the thirty-four cases. Therefore, the classification procedures were quite complex, and it seems inappropriate to describe them in detail here.

The essential classes, covering a large majority of comparisons and decisions, were those indicating strong or moderate support for the hypothesis, and those indicating strong support for the alternative hypothesis of cost minimization. Because it was assumed only that the discussed alternative technique put product advantage at risk—not that it made it impossible to achieve—a comparison was judged to provide "strong" support for the hypothesis only when it revealed a positive NPB that was 2.5 or more times greater than NTC, whether the latter was positive or negative. Moderate support was defined as a positive NPB, 1.5 to 2.5 times greater than NTC, again positive or negative.

On the other hand, strong support for a cost-minimization hypothesis covered cases where NTC was negative (indicating a cost saving in using the adopted technique) and 2.5 or more times greater than NPB, positive or negative.

Of the thirty-four comparisons, twenty-seven fell into these three classes: eighteen indicated strong support and six, moderate support, for the product-advantage hypothesis; and three indicated strong support for the alternative hypothesis of cost minimization.

The next analysis involved a procedure to give each decision equal weight. All the NPB–NTC comparisons pertinent to a given decision were assigned a combined weight of one. If, for example, there were two comparisons pertaining to a decision, one falling into one class and one into another, the decision would contribute 0.5 to the weight of each of the two classes. Using this procedure, the three essential classes accounted for 14.7 of a total of eighteen decision weights—8.9 indicating strong support for and 4.3 moderate support for the hypothesis, and 1.5 strong support for cost minimization.

These results, which indicate considerable support for the hypothesis of economically motivated pursuit of product advantage, were then tested for their sensitivity to 1) higher discount rates, 2) economies of scale that might have been overlooked as the advantage-concerned firm's alternative situation was projected on the basis of the operations of firms currently using the discussed alternative techniques, and 3) sharper downward trends in revenue/raw materials ratios than those projected in the actual situations of those firms.

All three of these factors tended against the hypothesis by lowering NPBs, raising NTCs, or both. But even when all three were admitted in combination, they did not affect the level of strong support for the hypothesis, although they did greatly erode the level of moderate support for it: the eighteen of the thirty-four comparisons and the 8.9 of the eighteen decision weights that indicated strong support for the hypothesis prior to the sensitivity analysis continued to do so; but the degree of moderate support dropped substantially, from six comparisons to three and from 4.3 decision weights to 1.7.

Thus, the sensitivity analysis shows uneven support for the hypothesis. There is a solid core of comparisons and decisions indicating strong support for the hypothesis that is not eroded by sensitivity analysis. But that core does not account for a majority of the eighteen decisions. Even the addition of decisions continuing to indicate moderate support in the face of sensitivity analysis does not yield large majorities in combined strong and moderate support: twenty-four of the thirty-four comparisons and 10.6 of the eighteen decision weights. Although this research firmly supported the first hypothesis, that firms choose technologies with the goal of achieving product advantage, the second hypothesis, postulating an economic motivation for that concern, receives only moderate overall support. It should also be noted, however, that there was very little support for the alternative hypothesis of cost minimization—only three of the thirty-four comparisons and 1.5 of the eighteen decision weights.

DISCUSSION

Taken as a whole, the research reported here indicates that the competitive model of homogeneous products has little applicability to Indonesian industrial conditions. Widespread product heterogeneity was found, and the primary concern of firms in adopting techniques was to achieve some product advantage. Moreover, the vast majority (thirty-two of thirty-four) of "net product-associated benefits" evaluated were positive. Thus, product advantage was generally translated into an economic benefit, at least in the product markets. In these circumstances, one should not expect cost minimization to be the dominant behavior. Although much of the discussion of industrialization in less developed countries has implicitly assumed that the overriding concern is to cut costs, the results of this study do not indicate much support for a hypothesis that cost minimization determines technology choice.

Sensitivity analysis showed somewhat uneven support for the notion that firms' pursuit of product advantage is economically motivated. This slightly puzzling result may be partly explained by the incidence of branding. It may be that multinational firms, encouraged by a tradition of exporting high-quality products at high prices from developed countries to Indonesia, try to reproduce the quality of previous imports by adopting techniques that minimize the risks to such import substitution. And indigenous trading firms turned manufacturers may also have had a tradition of importing and selling such products, and hence may emphasize product heterogeneity in their choice of production techniques. Both types of firms may expect that high levels of product-associated benefits would outweigh any concomitant technique-associated cost increment. But it may be that these hopes are fulfilled only when product advantages can be converted into high NPBs by the establishment of well-known brand names, which lower the price elasticities of product demand.

There is in fact some relation between strong branding and strong support for the hypothesis of an economic motive for the pursuit of product advantage. Branding was not well established in printing, weaving, and textile printing—industries in which support for the hypothesis was particularly weak—and in weaving there was considerable support for cost minimization. Weaving is a competitive industry by Indonesian standards, with several firms producing in each quality segment, so the evidence from this industry tends to corroborate Yeoman's finding that high product-price elasticities of demand give manufacturers a significant incentive to adopt techniques reflecting relative factor prices.[4]

However, the incidence of branding cannot fully explain the uneven pattern of support for the hypothesis. In two of the industries with strong branding—paint and soap—all the decisions indicated strong support for the hypothesis. However, branding was equally well established in cigarettes and flashlight batteries, where only one-third of the decisions indicated strong support.

It is useful to look more closely at the purposes underlying the decisions offering, at most, moderate support for the hypothesis, of which there were nine. Three were in cigarettes, made by a multinational firm whose primary and stated purpose in producing to high quality standards was not to reap short-run profit, but to establish itself in a substantial, growing national market. Its long-run strategic motive is evidently not well captured by the medium-term discounted-cash-flow methods underlying the analysis of economic motivation in this research. Three more of these decisions were in the printing and textile-printing industries. These were fairly obvious cases of miscalculation—adoptions of unnecessarily high-cost techniques due to inexperience. The firms that had made these choices had either corrected their mistakes in subsequent adoptions or stated their intention of doing so as soon as the opportunity for replacement investment arose. The remaining three decisions, in flashlight batteries and weaving, have no special circumstances bearing on their low level of support for the hypothesis.

Thus, it appears that in two-thirds of the decisions offering at most moderate support for the hypothesis, the economic motivation may have been stronger than was recognized by the study's tests. This finding increases the consistency and overall level of support lent by the research to a finding of an economic motivation for decisions, though it also demonstrates that such motivation is not always well captured by medium-term financial analysis or by the initial decisions of possibly inexperienced firms.

The issue of firms' experience prompts a final question pertinent to the general field of research encompassed by this volume. As multinational firms gain experience in producing a particular product line in additional less-developed countries, do they tend to make fewer economic miscalculations in choosing quality and production technique? The answer must be left for future research.

5
Choice of Technology in Thailand

DONALD J. LECRAW

By the late 1970s, several studies beyond those of Yeoman, Wells, and Keddie in this volume had confirmed that variables other than factor prices influence the choice of technology. Stobaugh et al. found in 1976 that U.S.–based multinationals had a strong propensity to purchase equipment from their home country. They offered as explanation the reduction of risks and of search and information costs. In 1979, Morley and Smith, in a study of foreign firms in Brazil, supported the latter explanation;

This study was partially funded by the Centre for International Business Studies, School of Business Administration, The University of Western Ontario. It is based in part on the author's "Choice of Technology in Low-Wage Countries," Ph.D. dissertation, Harvard University, 1976.

85

they also concluded that firms in Brazil were not forced to use an appropriate technology, since the competitive environment there allowed inefficient firms to exist. (In an earlier paper Bergsman had similarly concluded that the competitive behavior of firms in Brazil allowed them a wide range of nonoptimizing behavior.) White, in 1976, found similar evidence with respect to firms in Pakistan.[1]

Despite these studies, most of the research by economists proceeded on the assumption that firms select technology to minimize costs and maximize profits, so that relative factor prices alone determine what technology is optimal and will be chosen by firms. In their work, economists made extensive use of the concept of a production function, that is, the technological relationships between the output of a producing unit and the inputs of factors of production used by that unit. The producing unit can be defined at various levels of aggregation—a factory, a company, an industry within a country, or an entire national economy. A production function can be used to describe, for a given level of output, the substitution possibilities among the factors of production. A low elasticity of substitution means that there are few possibilities to substitute one factor of production for another, say, labor for capital. A high elasticity of substitution, on the other hand, means that labor can be readily substituted for capital. If firms in the manufacturing sector of countries with low wages and high capital costs cannot or do not substitute labor for capital, industrialization will not create as many jobs in their economies as are needed to reduce unemployment and underemployment. Hence the elasticity of substitution and the factors that influence firms' choice of technology are important determinants of the success of economic development programs that emphasize growth through industrialization.

THE THEORY OF THE PRODUCTION FUNCTION

Production functions are usually stated as mathematical relationships between factor inputs and output. To model the real world using the relatively few variables that can be readily handled mathematically, a number of simplifying assumptions are required. Considerable attention has been focused on the simplifying assumptions that might be appropriate in describing the elasticity of substitution. A convenient and often-used assumption is that, regardless of level of output or degree of technical progress, the elasticity of substitution remains constant over all output levels and capital-to-labor ratios—hence the term, Constant Elasticity of Substitution, or CES, production function.[2] A CES production function for output PP is displayed in figure 5-1. Output PP can be produced using any combination of labor and capital along the curve PP. If the firm pays wages w and has capital costs r, then its costs are minimized if it operates at point C, using L_C units of labor and K_C units of capital. C is the point at which PP has the slope $-w/r$. If the firm were to operate at some other point along PP (for example, B) or to the right of PP (for example, at A), its costs would be higher than if it operated at point C. Hence economic theory predicts that a firm facing a CES production function, PP, producing output with wages w and capital costs r, will operate at point C.

Since the seminal work by Arrow, Chenery, Minhas, and Solow in 1961,[3] many studies have estimated the elasticity of substitution of some form of the CES production function. The estimated values of the elasticity of substitution for the same industry in different countries have differed widely across studies, as have estimates of the elasticity of substitution using different data sets and estimation techniques.[4] This wide variation could have several possible causes:

1. The CES production function itself may not be correct, i.e., elasticities of substitution may depend on capital-labor ratios, scale, or degree of technical progress and hence may vary among countries that differ in these respects.[5]
2. Errors may arise in using undifferentiated capital and undifferentiated labor as the sole inputs, and there may be problems in measuring capital and labor inputs, however defined.
3. The use of *industry-level* data, at whatever level of aggregation, may mix fundamentally different products, e.g., digital and mechanical watches.
4. The use of aggregate data raises other questions: it mixes *ex ante* and *ex post* substitution in response to changing factor prices over time, and it may give spurious results if changing capacity utilization is not taken into account.[6]

Fig. 5–1.   Choice of technology for a firm with factor costs w, r.

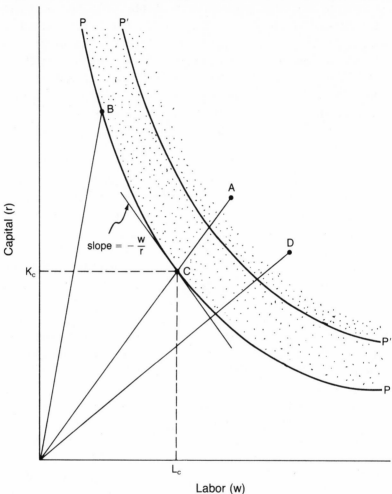

5.   The estimates of the elasticity of substitution can be sensitive to small changes in the data set.[7]

These difficulties led Morawetz to call for a more detailed, firm-by-firm examination of the range of technology available to firms rather than the use of aggregate data to estimate a production function.[8]

This chapter combines both approaches in that it uses data from individual firms to estimate the different production functions faced by firms in selected light manufacturing industries in Thailand. Then, in the spirit of the work of Yeoman, Wells, and Keddie, the assumptions of profit-maximizing behavior, risk neutrality, and perfect information are relaxed, and an effort is made to explain the choice of technology as a function of not only the factor prices (the cost of labor and capital) faced by the firm, but also its competitive environment, its strategy, its ownership, and certain characteristics of its management. The theoretical constructs of Simon, Leibenstein, and McCain about behavior *within* firms are also employed to help explain why managers do not always choose the profit-maximizing technology.

Simon used the term "satisficing" to describe one type of behavior he observed in managers and workers.[9] He found that so long as performance was above some satisfactory minimal level, neither the employees nor the owners of firms took steps to bring the firm's performance up to the unknown optimal level. This behavior is the result of two factors: in an uncertain, complex, changing world, optimal performance is difficult or impossible to define, measure, and motivate; and goals for the future, in terms of costs, revenues, profits, and growth, tend to be set in relation to achievements in the past. If optimal performance were achieved in the past, changing conditions might make the goals set in relation to the optimum unobtainable in the future. Thus managers have an incentive to build slack into the system; the slack can be squeezed out and performance improved if performance goals are threatened by changing conditions. The choice of technology is one of the decision variables that managers might use to build some slack into the firm's operations, to be reduced if profit or cost targets were in jeopardy.

Leibenstein developed the theory of X-Inefficiency to describe the effects of the way a firm is organized and operated to achieve its goals.[10] He concluded that because of imperfect information, lack of control, and incomplete contracts, there can be "inert areas" within a firm that permit individuals to behave in suboptimal, nonmaximizing ways, given the goals of the firm's owners. External factors—competition and pressures on profits—may push owners and managers to gather further information on the activities and decisions within their firm in order to reduce these inert areas and return the firm to an acceptable level of profits. McCain saw low profits as both an incentive to change and a signal that some change is necessary.[11] If these theoretical constructs are correct, firms that projected high profits at the time of their initial investment might operate with a high level of X-Inefficiency.[12] One of the forms this X-Inefficiency could take might be the choice of inappropriate technology. In competitive industries (where profits are usually lower), low projected profits might alert firms that their initial choice of technology will lead to high-cost production and thus give them the incentive to search for and use more appropriate technology, as indeed Yeoman found.

Wells postulated that the managers of firms, in their choice of technology, trade off profits ("economic man") against the satisfactions of using modern, fast, capital-intensive technology ("engineering man").[13] For the firms in Wells's sample in Indonesia, this engineering-man behavior was motivated partially by a desire to reduce risk of unacceptable quality, and partially by the belief, often a mistaken one, that the most technically advanced machines in an engineering sense were also the most economically efficient ones. Firms managed by engineering men would then tend to use a more capital-intensive, less appropriate, technology than firms managed by economic men, who would be less willing to sacrifice profits for engineering-man satisfactions. Wells suggested that the small number of firms in each industry, together with high import tariffs, reduced competitive pressures in Indonesia. Firms were not forced to use economically efficient technologies, and their managers and owners could indulge their engineering-man propensities instead of minimizing costs in order to maximize profits.

Another factor, however, may lead engineers to choose *more* appropriate technology than might nonengineers: engineers may be more aware than nonengineers of the range of technologies available to the firm and of the cost implications of using each technology; in addition, engineers may be less likely to treat the production process as a black box that converts raw materials into finished products and more likely to see it as a series of processes for each of which a range of technologies is available. The ultimate impact of engineers on the choice of technology cannot be known a priori.

Risk aversion may also influence the choice of technology. If a firm's managers fear that the use of a cost-minimizing technology will expose the firm to the risks of technological failure, or loss of quality or quality control, or risks associated with labor in a labor-intensive production process, they may choose a higher-cost technology if it reduces these risks. Keddie's study in Indonesia found that although managers often did not choose the lowest-cost technology, their decisions were not necessarily noneconomic.[14] Indeed, their primary concern seemed to be to make a product with a premium quality or to gain some other objective advantage that would allow the firm to secure a price premium large enough to justify the extra manufacturing costs.

METHODOLOGY

The impact of these variables on the choice of technology was tested by examining the choice of technology of a sample of 400 firms in twelve five-digit SIC (Standard Industrial Classification) industries in the light manufacturing sector in Thailand. These firms had all made substantial new investments in Thailand over the period 1962–1974. For 200 of the firms, an extensive questionnaire was administered during the course of an interview; a shorter questionnaire was filled out for the remaining 200 firms.[15] Data were collected not only on the economic characteristics of the firm—capital and labor employed, projected profit, number of shifts, output, and factor costs—but also on variables such as ownership, the education and experience of the firm's managers, their perception of various risks attached to their operations, and the firm's strategy. Data were collected for the projected value of the economic variables at the time of the investment and for their actual value if the firm had already begun full-scale production.

Even within these twelve narrowly defined industries, the range of technology chosen was very great: on average the capital-labor ratio of the most capital-intensive firm in an industry was 3.7 times greater than that of the least capital-intensive firm (see table 5–1). Clearly, a wide range of technology was available to firms in Thailand in these light manufacturing industries.[16] A comparison of actual capital-labor ratios with those projected by firms at the time of technology choices showed small differences for equipment but larger discrepancies for buildings.[17]

A CES production function was estimated for each of the twelve

TABLE 5-1

THE AVERAGE AND RANGE OF CAPITAL-LABOR RATIOS FOR FIRMS IN EACH OF 12 FIVE-DIGIT INDUSTRIES, THAILAND

SIC Industry No.	Name	No. of Firms	Average Capital-Labor Ratio (Units of 10,000 baht/Worker)	Range Capital-Labor Ratio (Units of 10,000 baht/Worker)
20	Food	35	8.3	3.3–12.7
21	Tobacco	30	11.3	3.2–10.5
23	Apparel	35	13.1	4.1–13.8
24	Wood products	32	7.5	3.5–11.1
26	Paper	32	6.7	2.7– 8.6
27	Publishing	38	18.1	6.2–19.4
30	Rubber	32	12.2	6.3–22.1
31	Leather	30	12.1	5.7–18.5
33	Metals	34	15.1	4.5–15.1
35	Nonelectrical machinery	38	10.2	2.2– 9.5
36	Electrical machinery	30	8.5	3.8–12.4
37	Transportation equipment	34	7.2	2.1– 7.3
	All manufacturing	400	10.9	2.1–22.1

NOTE: Industries are displayed at the two-digit rather than five-digit level to preserve the anonymity of the firms in the sample.

industries (see appendix to this chapter). For estimation of the production function of each of the twelve industries, the explained variable was value added by the firms in the industry, and the explanatory variables were combinations of labor, capital, time, and factors that depict returns to scale and technical change. Most work on the CES production function has assumed that product markets are perfectly competitive and that the firms employ the most efficient combinations of capital and labor. With these assumptions, an equation depicting the production function can be manipulated to yield a subsequent equation that can be estimated by linear techniques—a useful result that greatly aids the statistical analysis. But since these assumptions are the subject of investigation in this study, this method could not be used here. Instead, the equation for the production function for each industry was estimated by searching for those values of the explanatory variables that maximize the correlation of the actual and predicted values of the industry's value added. The strengths and weaknesses of this method, as well as those of other estimation techniques, were described by Feldstein.[18]

Since the data used were gathered on a firm-by-firm basis in narrowly defined industries and were taken from investment *projections,* several of the problems that ordinarily plague the estimation of a CES production function were avoided. Because labor was defined as the number of workers per shift, the coefficients estimated were not affected by different or changing levels of capacity utilization.[19] Since the data were *ex ante,* firms had complete freedom to choose their technology in relation to current and future factor prices and were not trying to alter, in response to changing factor prices, the capital-labor ratio of technology already in place; i.e., the data did not mix *ex post* and *ex ante* elasticities of substitution or technical progress.

A substantial methodological problem remained, however. When a production function is estimated, it usually is assumed that the firms are operating at the optimal points, but, as noted earlier, there may be considerable inefficiency in a firm's choice of technology.[20] If so, then the production function that is estimated will be some "average" production function, not the equation with the optimal efficiency. If the firms in the industry operated at the scattered points in figure 5–1, then the usual estimating procedures would derive the production function P'P', not PP as required. Two different mathematical procedures were used to try to surmount this problem and to obtain the efficient production function. Both approaches gave similar results for the deviations from optimal technology; one was chosen to provide the basic equations assumed to be the efficient CES production function for the individual industries. (See Appendix to this chapter for a detailed discussion of one of these procedures.)

Measures of Efficiency

By comparing this estimate of the efficient production function for an industry, PP, with the actual choice of technology used by individual firms within the industry, one can assess the efficiency of the technology each firm chose.[21] Two types of inefficiency in the choice of technology have been defined.[22] A technology is said to be *price* inefficient if it is the most efficient technology for any given capital-to-labor ratio other than the one faced by the firm using the technology. A technology is said to be *technically* inefficient if it uses the proper capital-to-labor ratio but uses too much capital and labor.[23]

Three distinct (though interrelated) measures of efficiency were used. The first is a measure of the *technical* efficiency of the firm's chosen technology; it is defined as the ratio of actual value added by the firm to the value added if it had used the minimum-cost technology. The second is a measure of *price* efficiency; it compares the actual capital-to-labor ratio used by the firm to the one that would have minimized costs for the firm. The third measure used here merges technical and price efficiency; it compares the *cost of operating* with the chosen technology (based on actual use of capital and labor at actual factor costs) with the theoretical minimum-cost technology.

Firms in Thailand showed both technical and price inefficiency (see table 5A–2 in Appendix), even though the decisions examined were for new investments, and the firms were not locked into inappropriate technologies by past choices. For example, the *technical* efficiencies of individual firms averaged 77 percent, with a range of 44 to 147 percent. *Price* efficiency, which is a measure of how much the actual capital-to-labor ratio differed from the optimum, averaged 63 percent, with a range of −48 percent to +271 percent. (A minus sign indicates that the actual capital-to-labor ratio used was lower than the optimum; a positive value means the actual was higher than the optimum.) The third measure, which merges technical and price efficiency, indicates that actual operating costs exceeded the minimum operating costs by 53 percent, with a range from − 22 percent to + 131 percent.

In contrast, Timmer and Aigner et al. found very little technical inefficiency in agriculture or industry in the United States.[24] Meller, on the other hand, found that three-quarters of the Chilean firms in his sample operated at less than half the efficiency of the most efficient firms in the industry.[25]

WHY FIRMS SELECTED INAPPROPRIATE TECHNOLOGY

What considerations led these firms to their inappropriate technology choices? The most obvious factor (and one well within the economist's neoclassical framework) is risk. If the owners or managers of a firm think that an appropriate technology is risky, they may accept the higher manufacturing costs of using an inappropriate technology so as to reduce their risk of operations.

Quantifying risk perception is a difficult task at best, and one that does not admit of any precision. As a first step, the managers responsible for the choice of technology in each firm in the sample were asked to rate, on a scale of 1 (low risk) to 10 (high risk), the various risks associated with operations in Thailand: general business risk, risk of expropriation, production disruption or cost increases (labor strikes, wage increase, input shortages), variable sales, capital unavailability, variable quality, low quality, use of labor-intensive technology, and shift work. The perceptions of risk were quite consistent for different managers in the same firm. A manager's risk perception on these dimensions might also be a function of his education (engineering or general), his experience in operating in a less developed country (LDC), the ownership of the firm (foreign or local), and the strategy of the firm (marketing or production orientation). Regression analysis indicated that these variables explained about half of the variation in managers' perception of overall risk.[26]

One way to decrease the perceived risk of technological failure while reducing search, information, and training costs would be to source technology from the home country of the firm. The findings of

Stobaugh et al., in their study of U.S.–based multinationals, and of Morley and Smith in their study of Brazil, were supported by the data for firms in Thailand. Foreign-owned firms tended to import their technology from their home countries. Japanese firms obtained 80 percent of their equipment at home, European firms 57 percent, and U.S. firms 51 percent. Interestingly enough, while firms based in other LDCs obtained 25 percent of their equipment locally, Thai-owned firms obtained only 13 percent locally.[27] Such a pattern might indicate that firms made only a limited search before choosing their technology. Managers were more interested in commencing operations as soon as possible, to take advantage of the high profits available in Thailand during this period, than in spending additional time and managerial resources in finding the lowest-cost, most appropriate technology. This supply pattern does not necessarily imply that foreign-owned firms chose a less appropriate technology than locally owned ones, since labor costs were higher for foreign-owned firms and appropriate technology may have been available at home.

The theory of X-Inefficiency suggests several factors that might influence the choice of technology. The lower the profits projected by the firm, the more pressure would be exerted within the firm's organizational structure to reduce the inert areas of nonoptimizing managerial behavior, one of which might be choice of inappropriate technology. Relatively low yet still acceptable projected profits would reduce the latitude for engineering-man behavior and could limit managers' ability to reduce their personal risk through their choice of technology.

Lower profits are often a function of increasing competition within the industry. The number of firms in an industry is one measure of its competitiveness. (A better measure is the level of concentration in the industry corrected for import penetration, but this figure was unavailable and impossible to calculate for each industry and each year.[28]) McCain concluded that as the number of firms in an industry increases, the "free pool of information" about alternative technologies also increases, and so new entrants will use more appropriate technology.[29] McCain also concluded that if the firm is owner-managed, the owner has a more direct access to information flows within the firm and can reduce the extent of its inert areas and force managers to choose more appropriate technology.

A statistical analysis of two hundred firms for which adequate data were available was used to determine what variables might explain variations in the firms' deviations from appropriate technology (see Appendix). Perceived risk, including risk of technological failure and unacceptable output quality, had a strong influence on the choice of technology. This result supports Keddie's conclusion based on Indonesian data (see Chapter 4). Increased perceived risk of technological failure, for example, reduced the technical efficiency of the technology employed, increased the capital-labor ratio, and increased operating costs. (The coefficients were significant at the 5 percent level for all three measures. See Appendix.) The firms seemed to be willing to employ older, known, less technically efficient technology in the uncertain environment in Thailand to reduce the risk of technological failure and unacceptable quality. And managers chose an excessively capital-intensive technology in the hope of increasing their product's quality or reducing the risk of variation in quality.[30]

Indeed, when managers were asked to rank the quality of their competitors' products, the level of quality was rated lower, and variability higher, for highly labor-intensive firms. On the other hand, the levels of quality and of variation in quality were rated the same for firms that used either capital-intensive or intermediate technologies (see table 5–2). It would appear that firms could use an intermediate technology without a sacrifice in quality. The key differences between capital-intensive production and production using intermediate tech-

TABLE 5–2

QUALITY OF OUTPUT AND CAPITAL INTENSITY, 200 FIRMS, THAILAND

QUALITY	CAPITAL INTENSITY		
	Labor-Intensive	Intermediate	Capital-Intensive
Average quality (1 = high, 5 = low)	3.5	2.1	2.0
Standard deviation of level of quality	±.8	±.6	±.5
Variation in quality, average (1 = uniform, 5 = highly variable)	4.1	1.2	1.3
Standard deviation of variation in quality	±1.1	±.3	±.4

SOURCE: Field interviews.

nology were often not in the core processing and production units, but in the peripheral activities of handling raw materials and finished output, where they had little impact on the quality of the output. For the most labor-intensive technologies, however, substantial substitution of labor for capital did occur in the core processing and production units, with a consequent fall-off and wider variation in product quality.

The data for this study were gathered between early 1973 and late 1974. Although the labor force in Thailand had previously been quite docile, the revolution in October 1973 was followed by a period of considerable labor strife. The managers of twenty-five of the firms that had completed the questionnaire before October 1973 were asked to fill out the sections on risk again after these disturbances. There was a substantial (and significant) increase in their perceptions of risk, particularly risk associated with labor. Asked how they had changed their future investment plans, most managers replied that they would like to install more capital-intensive equipment. Firms that had previously chosen capital-intensive technologies perceived the smallest increase in risk and planned to change their technologies the least. "Inappropriate," capital-intensive technology was one means by which firms were able to reduce their perceived risk of labor problems.

The level of competition also had a strong influence on the firms' choice of technology. Firms that projected high profits, for example, tended to choose significantly less appropriate technology than firms whose projected profits were low. This result supports Wells's observation that when high tariffs and a concentrated industrial structure keep competitive pressure low, firms are more likely to choose inappropriate technology.[31] As competitive pressure—as measured by the number of firms in an industry—increased, firms tended to choose more appropriate technology in their new investments. The number of firms in an industry had an impact on the choice of technology even after account was taken of profitability, providing support for the hypothesis that an increased number of firms in an industry augments the pool of available information about the different technologies available for production. When competition increased over time as more firms entered an industry, the firms that had originally chosen inappropriate technology often incurred losses while their more technologically efficient competitors prospered.

Firms that followed a strategy of product differentiation and branding and that had high advertising and selling costs were classified as "marketing oriented." These firms tended to choose less appropriate technology, possibly because their managers devoted their scarce resources to the more important marketing aspect of the firms' operations, and because production costs for these firms were a less important component of total costs. The managers of these firms also tended to perceive a higher risk in loss or variability of quality than did the managers of production-oriented firms. Firms that followed a strategy of competing on a price basis with low marketing costs and an undifferentiated product chose a more efficient technology than did marketing-oriented firms. These production-oriented firms needed low production costs in order to compete, because of the higher level of price competition as well as the fact that their production costs were a high proportion of total costs. (Both findings are consistent with Yeoman's work reported in Chapter 2.)

Most of the firms in the sample produced for the domestic market. Ten firms in five industries, however, exported a large percentage of their output as a final product, and five firms exported their output as an intermediate product in a multinational sourcing network. The firms that exported final products were production-oriented; they tended to use even more appropriate technology than other production-oriented firms, since their products had to compete in international markets on the basis of price. By comparison, the firms that exported intermediate products to other units in a multinational enterprise operated at the extremely capital-intensive end of the range of available technology, since they perceived a high risk of variation in quality and disruptions due to labor problems, and were willing to increase costs to reduce these risks.

This analysis does not support the hypothesis that ownership affects the choice of technology. Foreign-owned firms did not seem to choose less appropriate technology than domestically owned firms. Operating costs, for example, were not significantly different for the two sets of firms. True, as might be expected, technical efficiency was slightly higher for the foreign-owned firms and price efficiency was slightly lower, but neither was significant at the 90 percent confidence level.

Nor did owner-managed firms seem to choose a more appropriate technology than nonowner-managed firms, even though owner-managers might be expected to have more direct access to information flows within the firm. One possible explanation for this unexpected result emerged during the interviews. Owner-managers were able to make a more explicit trade-off between costs and engineering-man satisfactions than other managers. Although nonowners also enjoyed using new, modern, capital-intensive technology, they were evaluated largely on performance (profits) and so were less willing to indulge in engineering-man behavior because it was not valued by the absent owner.

Finally, the last set of variables explored—the experience of the managers—did seem to affect the choice of technology. Firms whose managers had experience in operating in low-wage countries tended to use more appropriate technology than did firms whose managers had little experience. This effect of managerial experience applied to both foreign and Thai firms.[32]

Firms managed by engineers tended to choose more appropriate technology than did those managed by nonengineers. This finding does not imply that engineering-man behavior is unknown in engineers. Any such behavior, however, was outweighed by the engineers' broader knowledge of the available technologies and their ability to break the production process into several stages and to calculate the cost implications of various choices of technology at each stage. This conclusion is similar to that of Pack for firms in Kenya.[33]

Clearly, then, the choice of technology in low-wage countries depends significantly on factors beyond the relative costs of capital and labor. These findings represent a sharp departure from most of the traditional work of economists and give strong support to more recent work emphasizing reduction of risk, lack of competition, firm strategy, and the cost and availability of information as major determinants of technology choice. These ideas have been extended and tested, and their impact on the choice of technology has been quantified; furthermore, other variables have been added to the analyses. Because of the econometric difficulties described in this chapter, one should be cautious in interpreting the estimated magnitudes of the impact of individual factors. Many of the factors influence the choice of technology in complex ways that may not be picked up in a statistical analysis. They may also stand as proxies for other unobserved (and possibly unobservable) variables. Nonetheless, in spite of these formidable methodological problems, these factors accounted for much of the deviation of firms from their cost-minimizing technology.

The influence of risk, competition, firm strategy, and information costs on the choice of technology in low-wage countries carries important policy implications, both for the firms themselves and for the development policies of LDCs.

When firms in an LDC choose inappropriate technology for their operations, two corrective measures are usually suggested: expand the range of technology available to the firms, and change the factor prices faced by the firm so that they reflect the true scarcities of capital and labor in the economy. These measures would indeed influence the choice of technology. As in the previous chapters, however, the analysis discussed here suggests that the competitive environment in LDCs allows such a degree of nonoptimizing behavior in the choice of technology that these measures will have less effect than hoped. Measures that reduce the perceived risk of investment in the use of appropriate technology in LDCs and that reduce the level of profits in noncompetitive industries might also significantly encourage the use of a more appropriate technology. Such measures might include the provision of information for prospective investors on appropriate technology and training in its use for managers, engineers, and workers; and efforts to make the environment more competitive—possibly through reduced tariffs on imports, relaxation of government barriers to the entry of new manufacturers, and incentives to export—so as to encourage more firms to compete vigorously on the basis of price. These measures would be more difficult to design and administer than ones that would simply change the factor-price ratio, but they are well within the administrative capabilities of the governments of LDCs.

APPENDIX

Section I: Statistical Estimates of CES Production Function for Four Hundred Firms in Twelve Five-Digit SIC Industries, Thailand

Part A: Equation of CES Production Function

$$Q = A \left[(e^{\lambda_L \cdot t} \cdot L)^{\frac{\sigma - 1}{\sigma}} + \alpha (e^{\lambda_K \cdot t} \cdot K)^{\frac{\sigma - 1}{\sigma}} \right]^{\frac{\gamma \sigma}{\sigma - 1}}$$

Where:

   Q = industry output, as measured by value added
   L = labor, defined as number of workers per shift
   K = capital stock, in units of 10,000 baht (deflated to take into account inflation in the price of the various components that made up the total capital stock)
   σ = elasticity of substitution
   γ = returns-to-scale factor
   t = time
   λ_L = technical change factor for labor
   λ_K = technical change factor for capital

Part B: Discussion of Estimation Procedures

The production isoquant is by definition the locus of the efficient combinations of inputs necessary to produce a given output. When the production function (the equation in Part A in this Appendix) is estimated, it is assumed that the only error is in the disturbance term. This error is generally thought to arise from variables omitted from the relationship or errors in measurement of the dependent variable. If firms do not operate along the production frontier, for whatever reason, the production function that is estimated will be some "average" production function, not the efficient frontier. This problem is illustrated in figure 5–1.

The production frontier is PP, and the estimated function is P^1P^1. For a given factor-price ratio, an efficient firm would choose its technology at C. Farrell defined two types of inefficiency in the choice of technology. A technology is *price*-inefficient if it is on the production frontier but the ratio of its marginal products of capital and labor is not equal to the ratio of their prices (point B, for example, in figure 5–1). A technology is *technically* inefficient if it is to the northeast of the production frontier (point A). At point D, the chosen technology is both price-inefficient and technically inefficient. For the purpose of this chapter, a firm that chooses technology at C is said to use "appropriate" technology; any other technology is said to be "inappropriate."

The problem is to estimate the equation in Part A of this Appendix (and hence PP) using data from firms that may have chosen inappropriate technology, i.e., both price-inefficient and technically inefficient firms. After the pioneering study of Farrell in 1957, not much work was done in this area until the early 1970s, probably because of the formidable theoretical and computational problems involved. But between 1971 and 1977, a series of articles attacked this problem, using a variety of techniques of increasing sophistication (see note 23). Because these techniques have so far been applied only to linear regression, the production function must be of a form that is linear or can be made linear by suitable transformations. Unfortunately, as already described, the CES production function cannot be reduced to a linear form without using strong assumptions about the behavior of the firms in the sample.

Two procedures were used to try to surmount this problem. Both gave similar results for the deviations from appropriate technology, which are calculated in the next section.

The first method used the CES production function equation (equation in Part A of this Appendix) and estimated it directly by the method suggested by Feldstein. No correlation was found between the sign of the residuals $\hat{Q} - Q$ and time, capital-intensity, or scale of operations. This gave some indirect support for a heroic assumption that the difference between the average production function and the best-use production function lay only in the intercept term A. The value for A that left 95 percent of data points to right, i.e., in the absolutely inefficient region, was then calculated. This procedure is roughly equivalent to that used by Timmer. The second method used a Cobb-Douglas $(C-D)$ production function logarithmic form:

$$\log Q = \log A + \lambda t + \alpha \log K + \alpha \log L$$

The techniques of Timmer and of Aigner, Lovell, and Schmidt were used to estimate the coefficients of time (t), labor (L), and capital (K). This procedure gave results similar to the others.

Part C: Results of Estimates

The values of the parameters of the CES production function equation estimated in this manner are given in table 5A–1. Notice that the elasticity of

TABLE 5A–1

CES PRODUCTION FUNCTION PARAMETER VALUES

SIC	σ	$\lambda_L - \lambda_K$	γ	$\bar{R}^2$
20	0.91	0.05	1.15	0.95
21	0.43*	0.04	1.02	0.77
23	0.55	0.11	1.43	0.87
24	0.85	0.12	0.95	0.91
26	0.92	0.02	0.87*	0.83
27	0.43	0.11	1.26	0.88
30	0.52	0.15	1.17	0.94
31	0.41*	0.17	1.14	0.91
33	0.62	0.11	1.02	0.90
35	0.87	0.09	1.03	0.85
36	1.05	0.06	1.11	0.92
37	0.97	0.09	1.07	0.82
All manufacturing	0.73	0.13	1.27	0.85

NOTE: All coefficients were significant at the 95 percent level *except* those starred (*).

substitution, σ, was quite high; there was technical progress of about 5 percent per year in a labor-augmenting direction; and there were increasing returns to scale (γ was greater than 1) for most industries.

Section II: Regression Analyses Explaining Why Firms Select Inappropriate Technology

Part A: Three Measures of Efficiency in Firms' Technology Choices (see table 5A–2)

$$D_1 = \frac{Q}{\hat{Q}} \text{ (technical efficiency)}$$

$$D_2 = \frac{K/L}{\hat{K}/\hat{L}} \text{ (price efficiency)}$$

$$D_3 = \frac{cK + wL}{c\hat{K} + w\hat{L}} \text{ (actual vs. cost-minimizing capital and labor expense)}$$

Where: Q = value added

K = capital stock

L = number of workers per shift

c = cost of employing capital equipment (cost of capital + depreciation + maintenance)

w = average wage

^ denotes value predicted on basis of production function, as opposed to actual value

Because the ratio of skilled workers and supervisors to regular production workers was not a function of capital intensity, w was treated as a composite wage. Both w and c varied over time.

In figure 5–1, for a firm operating at D, $D_1 \cong (OC/OD)$, $D_2 = (\text{slope } OD)/(\text{slope } OC) = (K^D/L^D)/(K^C/L^C)$, and $D_3 = (cK^D + wL^D)/(cK^C + wL^C)$.

If a firm chose appropriate technology, D_1, D_2, and D_3 would all equal 1.0.

Part B: Regression Equations Estimated

$$\left.\begin{array}{l} D_1 \\ |D_2 - 1| \\ D_3 \end{array}\right\} = a + bT + cR = dP + eN + fS + gM + hF + iE + jX + u$$

Where:

T = perceived risk of technological failure (scale 1 to 10)

R = perceived risk of unacceptable output quality (scale 1 to 10)

P = projected profits

N = number of firms in the industry at the time of the investment

S = strategy (0 = marketing, 1 = production)

M = management (1 = owner-managed, 0 = other)

F = percentage local ownership

E = engineering (1 = manager was an engineer, 0 = nonengineer)

X = experience (1 = manager was experienced in operating in LDCs, 0 = inexperienced)

Part C: Statistical Results

Note that an inappropriate technology results in a low D_1, a high $| D_2 - 1 |$, and a high D_3 (assuming that the inappropriateness with regard to D_2 is in the direction of excess capital intensity). Thus, one would expect the sizes of the coefficients in the equation for D_1 to be opposite to those in the equation for $D_2 - 1$ and D_3. (See table 5A–3.)

TABLE 5A-2

DEVIATIONS BY FIRMS IN THEIR CHOICE OF TECHNOLOGY FROM A COST-MINIMIZING TECHNOLOGY

SIC	D_1 Average	D_1 Standard Deviation	D_1 Range	D_2 Average	D_2 Standard Deviation	D_2 Range	D_3 Average	D_3 Standard Deviation	D_3 Range
20	.88	.06	.52–1.16	1.65	.37	3.50– .73	1.62	.19	2.21– .96
21	.74	.11	.56–1.23	2.02	.43	3.71– .38	1.73	.21	2.01– .89
23	.77	.09	.62–1.47	1.92	.37	2.95–1.18	1.61	.18	1.98– .78
24	.62	.10	.52–1.18	1.68	.41	2.21– .52	1.68	.21	2.17– .93
26	.73	.14	.47–1.14	1.75	.51	3.34– .65	1.41	.29	2.23– .89
27	.71	.13	.52–1.24	1.81	.54	3.07– .61	1.72	.31	2.27– .93
30	.68	.11	.50–1.23	1.72	.42	3.75– .52	1.73	.25	2.15– .92
31	.78	.06	.47–1.20	1.39	.21	2.92– .57	1.82	.15	2.31– .88
33	.78	.11	.47–1.25	1.87	.47	2.91– .83	1.45	.19	2.20– .87
35	.69	.10	.44–1.18	1.75	.23	1.93– .67	1.48	.21	2.42– .89
36	.72	.06	.57–1.14	1.25	.11	1.43– .77	1.51	.17	2.00– .94
37	.75	.07	.54–1.22	1.71	.29	2.29– .62	1.42	.18	2.05– .89
All manufacturing	.77	.05	.44–1.47	1.63	.43	3.71– .52	1.53	.12	2.31– .78

$$D_1 = \frac{Q \text{ actual}}{Q \text{ estimated}}$$

$$D_2 = \frac{(\text{Capital-labor ratio}) \text{ actual}}{(\text{Capital-labor ratio}) \text{ cost-minimizing}}$$

$$D_3 = \frac{(\text{capital} + \text{labor expense}) \text{ actual}}{(\text{capital} + \text{labor expense}) \text{ cost-minimizing}}$$

TABLE 5A–3

REGRESSION COEFFICIENTS EXPLAINING WHY FIRMS DEVIATE FROM OPTIMAL
TECHNOLOGY

EXPLANATORY VARIABLES	EXPLAINED VARIABLES		
	D_1	$D_2 - 1$	D_3
Constant	$.45^b$	1.7^b	1.8^b
	(2.3)	(3.1)	(3.7)
Risk variables:			
T × 10^3 (technology risk)	-2.3^b	$.57^b$	5.8^b
	(2.7)	(2.8)	(3.2)
R × 10^3 (quality risk)	-4.4^a	$.83^b$	6.2^b
	(2.2)	(2.9)	(5.3)
Competition variables:			
P (profits)	$-.021^b$	$.81^b$	4.5^b
	(3.5)	(2.9)	(4.8)
N (no. of firms)	$+.032^b$	$-.38^b$	$-.59^b$
	(3.3)	(3.5)	(2.2)
Strategy variables:			
S × 10^3 (strategy)	$.95^b$	$-.52^b$	-3.4^b
	(3.1)	(4.2)	(2.7)
Ownership variables:			
M × 10^3 (owner-manager)	-4.1^a	1.8^b	8.5^a
	(1.7)	(3.2)	(2.9)
F × 10^3 (percent local ownership)	$-.81^a$	$+3.8^a$	$+4.2$
	(2.8)	(2.4)	(.71)
Training and experience of manager variables			
E (engineer-manager)	$+.32^a$	$-.90^a$	-5.7^a
	(4.7)	(4.1)	(3.0)
X (experience in LDCs)	$+.072^a$	$-.83^b$	-5.2^b
	(2.1)	(2.7)	(2.4)
$\bar{R}^2$	.95	.97	.97
Number of firms	200	200	200

^{a} = significant at the 10 percent level
^{b} = significant at the 5 percent level
( ) indicates t-statistics

6

Technology Choice for Textiles and Paper Manufacture

MICHEL AMSALEM

In an attempt to understand technology choice in developing countries, this study seeks to answer four broad questions:

1. How wide a range of alternative technologies actually exists for the production of a given good, at a given scale of production, especially in the little-studied process industries? And to the extent that alternative technologies are possible, how different are their requirements for the factors of production?

This chapter is based on the author's "Technology Choice in Developing Countries: The Impact of Differences in Factor Costs," D.B.A. dissertation, Harvard Business School, 1978, from which the book, *Technology Choice in Developing Countries* (Cambridge, Mass.: MIT Press, 1983) is derived.

2. How does the technology that results in the minimum cost of production from the firm's point of view compare with the technology that results in the minimum cost of production from the country's point of view? That is, what impact do distortions in the cost of the factors of production have on the choice of technology?

3. How do the technologies chosen by firms compare with those that would have resulted in the lowest cost of production for the firms? And how do these technologies compare with those that are optimal for U.S. conditions?

4. When the cost-minimizing technology is not the one adopted by a firm, how well do the studies reported in the previous chapters explain that decision?

METHODOLOGY

This study examined firms in two industries—textiles (more specifically, the spinning and weaving of short cotton and synthetic fibers) and pulp and paper—in four developing countries: Colombia, Brazil, Indonesia, and the Philippines. Selected plants in the United States and Japan were also studied for comparison with the plants in the developing countries. In total, the study includes choices of technology made in nineteen textile facilities and in fourteen pulp and paper mills. Additional information about the technologies available for each of the processing steps in the two industries was obtained through interviews with equipment manufacturers and engineering and consulting firms.

A three-step methodology was followed: first, for each industry the amount of the factors of production required for the production of a given standard quantity of output was determined for each alternative technology available for each processing step in the manufacturing process. Second, for each processing step and for each firm, the production cost of each alternative technology was estimated on two bases: the market cost of the factors of production to the firm, and the social cost of these factors to the country in which the production facility was located. The production costs of these alternative technologies were then compared to determine which would minimize market costs and which would minimize social costs (that is, to identify the "market-optimum" technology and the "social-optimum" technology). The technology actually chosen by the firm was then compared with these.

Third, the propensity of a firm to minimize its cost of production by adapting its choice of technology to the local environment (that is, the "propensity to adapt") was measured by a three-step calculation:

1. The actual savings the firm realized was calculated by subtracting its actual cost of production from the cost of production using U.S. technology.
2. The potential savings that could have been realized was calculated by subtracting the cost of production using the market optimum technology from the cost of production using U.S. technology.
3. The actual savings was divided by the potential savings to determine the propensity to adapt.

Thus, if a firm actually chose the market-optimum technology for a certain processing step, its propensity to adapt would be equal to 1.0. Since such ratios are independent of the range of alternative technologies available for any given processing step, they make it possible to compare the level of adaptation in different processing steps, different companies, and different industries.

This methodology presents several advantages over the use of capital-labor ratios or labor-output ratios on which studies of technology choice have traditionally been based, in that it allows for

1. The evaluation of specific choices of technology against a set of *existing* alternatives rather than a theoretical production function that may not be relevant in practice;
2. The identification of optimal technologies from the point of view of the country as well as that of the firm;
3. The consideration of differences in the requirements of factors other than labor and capital;
4. The consideration of differences in the skill level and therefore the cost of labor required; and
5. Comparisons of behavior across processing steps, across firms, and across industries.

ALTERNATIVE TECHNOLOGIES

In both the textile and the pulp and paper industries, alternative technologies requiring significantly different quantities of capital and labor for the production of a given standard quantity of output were available in 1977 for most of the processing steps.[1] The alternatives available in the textile industry were especially numerous. Of the twelve processing steps studied, five had five or more technologies, three had four, two had three, one had two, and only one had just one available technology. Some steps had many more than five. For example, for the weaving step, ten different types of looms, each embodying a different technology, were available from equipment manufacturers.[2] In the pulp and paper industry, the choices of technology were more limited, with six of the sixteen processing steps studied having only three possible technologies, two steps having two, and the other eight having only one.

While neoclassical economic theory assumes that an infinite number of technological alternatives is available, their actual number is limited, and for some processing steps it is very small. Nevertheless, even for the cases where only a few alternatives are available, the relative quantities of capital and labor required are sufficiently different to make the choice of technology a relevant issue.

Table 6–1 shows the range of factor requirements for seven processing steps in textiles. For each factor type and each process step, the table shows the factor input required by the technology that uses the most of this factor as a multiple of the input required by the lowest-use technology. For example, the technology that employs the most labor in plucking uses 3.1 times as much total labor as the technology that employs the least total labor. But for semiskilled labor, the high-use technology employs 9.9 times as much as the low-use technology. The greatest variation in labor requirements shown in table 6–1 is for semiskilled labor in the carding step, where the ratio of most-to-least requirements is 29.9:1. For all processing steps, there was greater variability in the use of semiskilled labor than in the other three categories of labor.

The use of supervisory and skilled workers is of interest because they are perceived to be a relatively scarce resource in developing countries. Some authors have hypothesized that capital-intensive tech-

TABLE 6–1

Index of Factor Requirement of Technology Using the Most of that Factor in the Textile Industry[a]

Factor Requirements	Plucking	Scutching	Carding	Drawing	Winding	Pirn-Winding	Weaving[b]
Total labor requirements	3.1	11.5	14.3	2.2	8.2	4.5	11.5
Supervisory	1.9	3.3	5.7	1.6	3.4	2.3	3.2
Skilled	2.0	1.3	2.4	1.4	1.7	1.5	3.7
Semiskilled	9.9	c	29.9	4.7	18.0	5.4	19.1
Unskilled	2.0	19.0	23.7	2.0	1.0	1.0	2.5
Total investment requirements	1.6	1.8	1.8	1.5	3.4	2.9	3.9
Equipment	1.9	2.0	2.0	1.7	3.7	3.3	6.5
Buildings	1.0	1.4	1.5	2.7	1.5	c	1.3
Number of alternative technologies available	3	5	7	4	9	4	10

SOURCE: Michel A. Amsalem, *Technology Choice in Developing Countries* (Cambridge, Mass.: MIT Press, 1983), Table 2–4, p. 46.
[a]1.0 is the quantity of factor used by technology employing least amount of that factor.
[b]Handlooms not included because not typically used in industrial establishments.
[c]The technology using the least of this factor did not use any of it.

nologies are adopted in developing countries because they require less
skilled and supervisory labor than do labor-intensive technologies.[3]
This hypothesis was not supported in this study, as five of the seven
process steps in textiles contained capital-intensive technologies that
used absolutely greater amounts of skilled and supervisory labor than
did more labor-intensive technologies.[4]

Capital requirements vary much less than labor use. The widest
range was in the weaving step, where the technology using the most
investment required 3.9 times as much capital as the technology using
the least investment. For most processing steps there is greater variation in investment in equipment than in buildings.[5] In weaving, for
example, the technology using the most investment in buildings required only 1.3 times as much as the technology using the least, while
the technology using the most investment in equipment required 6.5
times as much as the technology using the least.

Although alternative technologies exist in both the textile industry and the pulp and paper industry, these alternatives take different
forms in each case. In the textile industry, the transformation process
is a relatively simple mechanical process with a limited number of
inputs; therefore, the alternative technologies are embodied in the
main piece of equipment. The pulp and paper industry, on the other
hand, is a chemical-process industry in which raw material is transformed through the action of chemicals, heat, and pressure. There is,
therefore, little room for manual intervention in the transformation
process, and the alternatives embodied in the main piece of equipment
are few. The trade-offs between capital cost and the cost of chemicals
and energy overshadow the possible trade-offs between capital cost
and labor cost. The reactions and flows taking place in these main
pieces of equipment, however, can be controlled in a variety of ways,
from the manual opening and closing of valves to computerized
monitoring and adjustments. The trade-offs between capital and labor
costs, therefore, are made in the choice of instrumentation and control
equipment rather than in the choice of the main processing equipment.
As shown in table 6–2, variations in three features of the control
process—the location of the control mechanism, the driving force of
the adjustment mechanism, and the source of control—result in seven
different technologies available for the control of the digesting step in
pulp and paper.

TABLE 6–2

LEVELS OF TECHNOLOGY AVAILABLE FOR THE CONTROL FUNCTION IN DIGESTING STEP IN
PULP AND PAPER INDUSTRY

LEVEL OF AVAILABLE TECHNOLOGY	FEATURES OF CONTROL PROCESS		
	Location of Control Mechanism	Driving Force of Adjustment Mechanism	Source of Control
No. 1	at adjustment mechanism	manual	human
No. 2	at adjustment mechanism	mechanical power	human
No. 3	at central instrument panel	mechanical power	human
No. 4	at central instrument panel	mechanical power	automatic; manually set valves
No. 5	at auxiliary instrument panel used for digesting process	mechanical power	automatic; manually set valves
No. 6	at auxiliary instrument panel used for one group of processes	mechanical power	automatic; manually set valves
No. 7	at auxiliary instrument panel used for one group of processes	mechanical power	automatic; automatically set valves

SOURCE: Michel A. Amsalem, *Technology Choice in Developing Countries,* Table 2–3, p. 43.

TECHNOLOGIES CHOSEN

Within the range of available technologies, the technology actually chosen by a particular firm was compared with the technology that would have resulted in the lowest cost of production for the firm (the market optimum) and with the technology that would have resulted in the lowest cost of production for the country (the social optimum). The technology best suited to conditions in the United States (the unadapted technology) was used as reference.

The Textile Industry

The market-optimum technology and the social-optimum technology for the developing countries were significantly different from the optimal technology for U.S. conditions. This result, shown in table 6–3, is not surprising given the differences between the United States and the developing countries in the cost of factors of production and the differences between alternative technologies in input requirements.

One hundred ten technology-choice decisions in developing countries were studied. (See table 6–4.) In forty-three cases (about 40 percent of the total), the firm opted for the technology that constituted the market optimum in the conditions under which it was operating. Technologies more capital intensive than the market optimum were

TABLE 6–3

U.S. TECHNOLOGY COMPARED WITH DEVELOPING COUNTRIES—MARKET-OPTIMUM AND SOCIAL-OPTIMUM TECHNOLOGY, IN TEXTILES

PROCESSING STEPS (and number of levels of technology available)	TECHNOLOGY LEVEL[a]		
	Market Optimum in United States	Developing Countries	
		Market Optimum	Social Optimum
Plucking (3 levels)	2	1 (10 firms) 2 (6 firms)	1 (all firms)
Scutching (5 levels)	5	1 (13 firms) 5 (3 firms)	1 (all firms)
Carding (7 levels)	5	2 (all firms)	2 (all firms)
Drawing (4 levels)	4	1 (4 firms) 2 (12 firms)	1 (15 firms) 2 (1 firm)
Winding (9 levels)	8	1 (7 firms) 2 (6 firms) 3 (3 firms)	1 (all firms)
Pirn-winding[b] (4 levels)	3	2 (10 firms) 3 (5 firms)	2 (all firms)
Weaving[b] (10 levels)	9	3 (8 firms) 4 (7 firms)	3 (all firms)

SOURCE: Michel A. Amsalem, *Technology Choice in Developing Countries,* derived from Table 4–1, p. 93.
[a]The numbers indicate level of technology, with 1 being lowest level (see Appendix for further information).
[b]Of the sixteen firms studied in developing countries, only fifteen had these processing steps.

TABLE 6–4

CHOICE OF TECHNOLOGY BY TEXTILE FIRMS IN DEVELOPING COUNTRIES AND THE IMPACT OF SUCH CHOICES ON COST, EMPLOYMENT, AND CAPITAL REQUIREMENTS[a]

	Technology Chosen				
	More Labor Intensive Than Market Optimum	Market Optimum	More Capital Intensive Than Market Optimum but Less Than U.S. Technology	U.S. Technology	More Capital Intensive Than U.S. Technology
Number of observations	17	43	35	12	3
(percent of total)	15%	39%	32%	11%	3%
Input requirements (over market-optimum requirements)					
Manufacturing costs	1.2	1.0	1.2	1.1	1.5
Employment	3.3	1.0	.7	.6	.9
Supervisory and skilled	1.4	1.0	.9	.9	1.2
Semiskilled and unskilled	11.3	1.0	.7	.6	.9
Capital requirements	.9	1.0	1.5	1.3	2.0
Equipment	.9	1.0	1.7	1.4	2.3
Buildings	1.3	1.0	1.0	.9	.3
Foreign Exchange	.9	1.0	2.3	1.4	4.6

SOURCE: Michel A. Amsalem, *Technology Choice in Developing Countries*, from Table 3–7, p. 78.
[a]1.0 is input requirement for market-optimum technology.

selected in fifty cases, or about 45 percent of the observations. In close to a third of these observations, the technology chosen was the one that would have been optimal under U.S. conditions or a technology even more capital intensive than that. Finally, in the remaining seventeen cases (15 percent of the observations), a technology more labor intensive than the market optimum was selected. Most of these cases of "overadaptation" were found in the three Colombian firms studied and can be explained by the fact that, although the cost of capital had been kept artificially low in that country, firms were faced with a scarcity of capital. Figure 6–1 illustrates the points made above in the specific case of the weaving step, identifying the chosen technology, the market optimum, and the social optimum for each of the eighteen plants studied.

In a majority of the cases, then, firms failed to adapt to market prices, substantially increasing their production costs. In the plants using technologies more capital intensive than the market optimum, manufacturing costs were found to be, on average, 10 to 50 percent higher than those that would have resulted from the use of the optimal technology. The impact on the use of labor, capital, and foreign exchange also was substantial. The plants that chose technology more capital intensive than the market-optimum technology created a third less employment, used 50 percent more capital, and twice as much foreign exchange as they would have if they had opted for the market-optimum technologies.

Despite distortions in the costs of labor, capital, and foreign exchange in all of the developing countries in which the firms were located, the social-optimum was identical to the market-optimum technology in more than half of the cases, because of the limited number of technology alternatives. In the cases where the social-optimum and the market-optimum technologies were different, the adoption of the social-optimum instead of the market-optimum technologies by the firms in the sample would have resulted, on average, in an 82 percent increase in employment, a 32 percent reduction in capital requirements, and an 18 percent reduction in foreign exchange requirements. These large differences in employment and capital use make the potential costs associated with factor price distortions important.

Overall, there was a large disparity between the chosen tech-

Fig. 6–1. Levels of chosen technology, the market-optimum, and the social-optimum, for weaving for 18 firms.

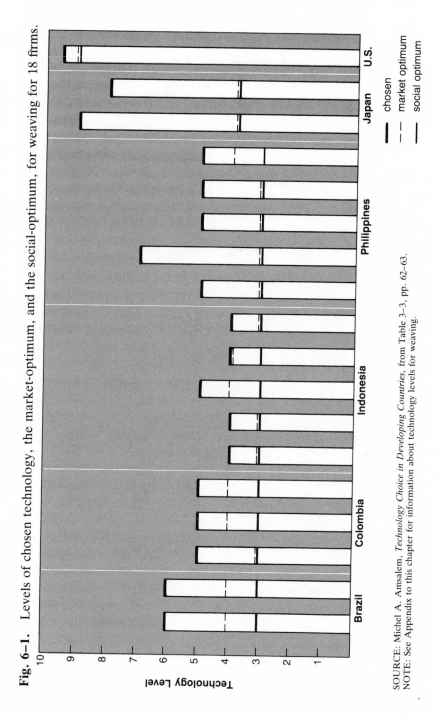

SOURCE: Michel A. Amsalem, *Technology Choice in Developing Countries*, from Table 3–3, pp. 62–63.
NOTE: See Appendix to this chapter for information about technology levels for weaving.

119

nologies and the social optimum in terms of capital use and employment. The chosen technologies used one-third more capital than social-optimum technologies, and provided only about half as many jobs.

The Pulp and Paper Industry

In the pulp and paper industry, the desirability of alternative technologies tends to be a question of scale rather than of the relative costs of the factors of production. The trade-off between capital and labor arises more in the choice of instrumentation and control equipment than in the choice of the processing equipment itself.

Table 6–5 illustrates the importance of scale. The cost of a paper-making machine, and the operating labor costs, increase less rapidly than the increase in output. Maintenance labor costs, on the other hand, increase about as fast as output. Overall costs are substantially lower if a single large machine (Machine E) replaces two smaller machines (Machine A).

The same pattern governs the choice of equipment for most of the processing steps in pulp and paper. It is less expensive to buy one piece of equipment that can process the mill's entire needs than two pieces of equipment each with half that capacity. At the same time, the number of operators needed for a given instrumentation and control technology tends to be proportional to the number of units rather than to the unit's size.

TABLE 6–5

INDEX OF MACHINE COST AND EMPLOYMENT ASSOCIATED WITH INCREASES IN PAPER MACHINE SIZE

	PAPER MACHINE				
	A	B	C	D	E
Actual production	100	123	154	160	200
Machine cost	100	108	119	122	133
Operating labor cost	100	102	105	109	112
Maintenance labor cost	100	118	144	155	193

SOURCE: Michel A. Amsalem, *Technology Choice in Developing Countries*, Table 6–2, p. 143.
NOTE: Estimates are based on the use of identical instrumentation and control technology in all cases.

Still, the mills studied showed some adaptation to factor costs. At similar scales of production, there were some differences in design between mills in developed and developing countries, principally in the choice of control technologies. Large mills in the United States, for example, consistently had more automated control technologies (for a total of twenty-one control variables for the six processing steps studied) than did large mills in developing countries. In the handling and transfer operations, more alternative technologies were available, and, on average, developing countries chose more labor-intensive technologies than did the United States. The technical feasibility and competitiveness of these labor-intensive handling technologies, however, decreased with increasing scale. For example, the handling operations in the wood yard were performed manually or with a pushcart in a developing country mill of 100 tons daily capacity. But another developing country mill, with 400 tons daily capacity, found such labor-intensive handling technologies impractical and had switched to moving belts in its wood-handling operations.

DETERMINANTS OF TECHNOLOGY CHOICE

The widespread adoption of technologies that involved production costs higher than those of the market-optimum technology suggests that these decisions were influenced by considerations other than the desire to minimize costs. Some insight into these considerations was given by interviews with the management and engineers of the firms studied and by comparisons of the technology choices made.

One important consideration was the availability and cost of information. To choose the best technology in a specific environment requires relatively detailed information about the input requirements of alternative technologies. In practice, the firms based their analyses on much scantier information. Furthermore, the information available to the firms tended to describe technologies at the more capital-intensive end of the spectrum, since information on technologies and equipment generally flows from developed to developing country, rather than from one developing country to another. But most of the labor-intensive technologies were available only from equipment suppliers in developing countries.[6] For example, the five most capital-intensive weaving technologies were only available from developed countries, while two of the three most labor-intensive could only be obtained from developing countries (see Appendix for definition). But all five lower-level looms were manufactured in developing countries under license from firms in developed countries.

In the pulp and paper industry, the unavailability of information on alternative technologies and equipment designs is due to the high cost of obtaining information rather than to some imperfection in information flows. Technologies must be adapted to the characteristics of the raw material used by each production unit, and the main pieces of equipment are custom made. Because of the costs involved in defining and evaluating each alternative technology, firms must minimize the number of alternatives evaluated. Firms' choices of technology were thus based on a limited range of alternative designs, and the main criterion used in deciding whether an alternative warranted detailed evaluation was whether it affected a factor whose control was critical to the mill's efficiency. Therefore, the alternatives that resulted in the use of a greater amount of labor ranked well below the alternatives that might lead to improvements in machine efficiency or wood yield, or to reductions in chemical and power consumption.

Considerations of risk entered the choice of technology in three different ways: first, through the evaluation of business and political risk; second, through the evaluation of risk associated with the use of the different factors of production; and third, through the desire to protect the competitive position of the firm.

In this study, the cost of the optimal technology was estimated on the basis of a cost of capital that reflected the evaluation of business and political risk by both investors and lenders to the project. As long as such an adjusted cost of capital was used, business and political risk should have had no further influence on the choice of technology. Foreign-owned firms, however, mentioned during the interviews that their decisions were influenced by a desire to minimize investment because of business and political risk in developing countries, even after these risks had been taken into account in the costing of capital.

This study also found evidence that firms tried to control risk associated with the use of the different factors of production. The risk associated with machines—the embodiment of capital—was considered to reside in the difficulty of getting spare parts and adequate technical assistance should a major technical problem arise. The firms tried to minimize this risk by choosing technology of simple, common design and by relying on well-known suppliers with large service networks.

One risk associated with the use of labor is the chance of labor unrest. Assuming that such risk increased more than proportionately with the size of the work force, firms tried to limit the number of workers employed and to prefer more automated technologies—when this could be done at a small cost. These tendencies, however, did not seem to increase with the size of the facilities, as might have seemed likely. Foreign-owned firms, which might be expected to be more sensitive to the risks associated with labor, did not choose more capital-intensive technologies than did local firms of similar characteristics and strategy. But this finding could be the result of offsetting factors working in opposite directions (e.g., the foreign firms' aim of limiting investment versus their wish to avoid having to manage a large labor force).

Another risk associated with the use of labor is the possibility of human error in the operation of the equipment. The cost of human error varies widely depending on the industry, the step in the production process, and the task to be carried out. It is higher in the pulp and paper industry than in the textile industry, and the choice of technology in the pulp and paper industry was correspondingly less sensitive to differences in the relative costs of labor and capital. Moreover, the risk of human error was mentioned repeatedly as the main factor in the choice of automatic controls for the pulp and paper plants. Because human error is more expensive in those steps of the production process that are crucial in determining the quality of the output or the efficiency of the plant, a higher degree of automation was found in these steps than in less crucial processing steps. Automated technologies were used more frequently in the processing tasks than in the material handling tasks, even if this resulted in higher overall costs for capital.

Deviating from the choices made by industry leaders constitutes a form of risk for smaller firms or newcomers to an industry. This factor had more influence on the choice of the equipment supplier than on the choice of technology as such. The "follow-the-leader" behavior of firms in choosing equipment suppliers may reflect the fact that certain suppliers had been prompted by large orders from industry leaders to expand their overseas service network. For similar reasons, foreign-owned firms often chose the same equipment suppliers as did their parent company, so as to take advantage of the established relations and of the accumulated pool of knowledge of headquarters facilities.[7]

The earlier chapters of this book suggest that firms that pursue a strategy of product differentiation adapt their technology less than others. Firms that do not differentiate their products face greater price competition and hence greater pressure to minimize costs, leading them to pay more attention to the possibility of adapting technology to local factor costs. This study has produced some evidence on this point. The twelve firms without a strategy of differentiation made a greater effort to choose labor-intensive technology than did the four firms with such a strategy. Measured on this basis, however, the difference does not appear to have been large (see table 6–6, part A). A greater difference appears when a more homogeneous group of firms is studied. For example, the three foreign-owned firms without a strategy of product differentiation chose a more labor-intensive technology than did the three differentiating foreign firms (table 6–6, part B). True, the three nondifferentiating firms had parents in developed countries, while the three differentiating firms had parents in developing countries, and the variation could reflect the type of country in which the firms were headquartered. The product differentiation

TABLE 6–6

PRODUCT STRATEGY COMPARED WITH CHOICE OF TECHNOLOGY

	STRATEGY OF PRODUCT DIFFERENTIATION	
	No	Yes
Part A. Number of Firms in Total	12	4
Choices of technology either at the market optimum or more labor intensive than optimum	57%	46%
Propensity to adapt	.79	.56
Part B. Number of Foreign Firms	3	3
Choices of technology either at the market optimum or more labor intensive than optimum	67%	32%
Propensity to adapt	.89	.45

SOURCE: Michel A. Amsalem, *Technology Choice in Developing Countries*, derived from Table 4–2, p. 100.

hypothesis seems more plausible, however, since, other things being equal, one would expect foreign firms headquartered in developing countries to use more labor-intensive technology than would firms from developed countries.

Within the sample of twelve firms that did not differentiate their products, three were owned by foreign firms headquartered in developed countries, seven by local private investors, and two by the local government. The non-product-differentiating firms with foreign ownership did not differ substantially in their degree of technology adaptation from privately owned local firms. Both groups, however, adapted substantially more than the firms owned by the local government (see table 6–7). Possibly the absence of a profit motivation in government-owned firms resulted in less incentive to adapt the technology to realize lower costs. The two government-owned firms in this study were managed by engineers, who appeared to choose equipment on technical rather than economic criteria.[8] The fact that the non-product-differentiating foreign-owned firms did not make technology choices substantially different from those of the privately owned local firms is contrary to the commonly accepted notion that foreign-owned firms adapt less than locally owned ones.

Finally, in several instances government policies not directly

TABLE 6–7

OWNERSHIP OF FACILITY COMPARED WITH CHOICE OF TECHNOLOGY, TWELVE FIRMS THAT DID NOT DIFFERENTIATE THEIR PRODUCT

	OWNERSHIP		
	Foreign Non-Product-Differentiating Firms	Local Private Investors	Local Government
Number of firms that did not differentiate their product	3	7	2
Choices of technology either at the market optimum or more labor intensive than optimum	67%	61%	29%
Propensity to adapt	.89	.81	.54

SOURCE: Michel A. Amsalem, *Technology Choice in Developing Countries,* derived from Table 4–2 p. 100.

reflected in the cost of the factors of production had an important impact on the choice of technology. Examples of such policies are the arbitrary allocation of scarce capital at subsidized rates, policies that favor local procurement of equipment, policies that increase the cost of financing imported equipment, and government tax incentives to attract foreign direct investment.

RELEVANCE OF THIS STUDY

This investigation has identified, for the textile industry and the pulp and paper industry, technologies that were available from equipment manufacturers but were not used or even considered by the firms in the sample. This type of analysis contrasts with the methodology used in most studies of technology choice, which compare the technologies chosen by different firms without reference to an independently established range of alternative technologies. Such studies often tend to conclude that there is little potential for adaptation in the mechanical industries and, because of the similarity of the choices made by firms of comparable scale in the chemical processing industries such as pulp and paper, that there is no potential for adaptation there. But the range of technologies available from equipment manufacturers suggests that the potential for adaptation is substantial in the textile industry and, although more limited, still significant for some operations in the pulp and paper industry. This study, therefore, points to the need for extended investigations into the range of alternative technologies available for the production of a given product. Indeed, the choice of appropriate technology could result in a lower manufacturing cost for the firm, and a substantial increase in employment and decrease in the use of capital and foreign exchange for the nation.

Furthermore, this study has developed a methodology for evaluating alternative technologies that is not limited to consideration of only two factors of production, as are methodologies based on capital-labor ratios. Through the use of this new methodology, differences in the requirements for labor skills, energy, material inputs, and buildings, as well as differences in the productivity of labor or in the efficiency of machine utilization, can be taken into consideration. The impact of these differences on the competitiveness of alternative technologies in a specific environment can be isolated. In addition, this methodology makes it possible to identify inappropriate choices of technology and to determine whether they are due to the relative use of various categories of investment, various categories of labor, or other inputs such as energy. Such an identification is difficult and imprecise when only capital-labor ratios are used. The tendency is often to assume that, in developing countries, the lower the capital-labor ratio the better, regardless of the market costs and the social costs of various factors of production.

The analysis of the different considerations that enter into the firms' choice of technology is based on too small a sample to permit strong, generalized conclusions. The results, however, are consistent with those reported in three of the previous chapters; and they also provide additional insights, especially as to the different types of risks perceived by managers. Moreover, the findings confirm the need for a closer examination of the choice of technology by state-owned enterprises.

APPENDIX

TECHNOLOGY LEVELS FOR WEAVING

In weaving, the *warp* yarn is held by the loom while the *weft* yarn is inserted across the warp in an interlacing fashion. The weft is inserted (pulled across the warp) either by means of a *shuttle,* which carries it back and forth across the warp, or, in shuttleless looms, by other devices. The shuttle contains a spool of wound yarn, or *cop,* which unwinds the weft as the shuttle is moved across the warp.

There are ten levels of loom technology embodied in equipment available from machinery suppliers. These technology levels have been ordered in such a way that, for a fixed output the capital cost of the technology increases and the labor requirements decrease as technology level rises.

For a further description, see Amsalem, *Technology Choice in Developing Countries,* pp. 34–37 and Appendix C, pp. 180–186.

7

Choice of Technology and Parastatal Firms

DAVID WILLIAMS

The previous chapters and most other studies have focused on technology choice in low-wage countries with market economies. Very little work has been done on technology choice in low-wage countries with state-controlled industrial sectors. Examination of investment decisions in such a country, Tanzania, reveals that decision makers in parastatal firms have chosen a wide range of technology. Although political leaders had clearly stated a preference for low-cost, labor-intensive, standardized technology, the technology selected did not meet the criteria implied by national

This chapter is based on the author's "National Planning and the Choice of Technology: The Case of Textiles in Tanzania," D.B.A. dissertation, Harvard Business School, 1976.

goals. The choices made could not be explained simply by a profit-maximizing motive or a preference for modern equipment. Rather, these choices seem to reflect the nature of the institutions involved in the decisions, the environment in which these institutions function, and the processes they have evolved for decision making. Since the national planning process failed to link the top-level statements of national objectives and the decisions affecting parastatals, technology choices were made through a process of bureaucratic interaction among the parastatal firms, planners, foreign investors, and other participants.

RANGE OF TECHNOLOGY IN TEXTILES

Initially, the study focused on the textile industry in Tanzania. In 1974, the cotton-spinning and -weaving industry consisted primarily of five major integrated mills (accounting for 85 percent of domestic consumption of woven cotton fabric), all subsidiaries of a parastatal holding company, the National Textile Corporation (TEXCO). Two mills had been established in 1967 as state firms; three had begun as private enterprises. At first glance, the range of technology used in these mills appeared very broad. One mill employed some 170 workers and Sh 1.6 million of fixed assets per million meters of finished cloth per year; another employed eighty workers and Sh 2.8 million for the same output. But the apparent differences could have resulted from operation of the mills at different levels of efficiency or from the manufacture of different mixes of products.

Detailed studies of the plants themselves were required to determine how broad the range of technology really was. To this end, data were collected from the mills and selected costs of production were brought to a standard basis for comparison. All data were adjusted to 1974 prices.

An examination of the two mills set up as state enterprises indicates no consistent criteria in the choice of technology. The more labor-intensive mill operated at some 80 to 90 percent of the cost of the more capital-intensive mill. If shadow prices were used to reflect the real costs of resources to the economy, it operated at some 70 to 80 percent of the social costs of the more capital-intensive one.[1]

In 1974, it was decided to expand the Tanzanian textile industry. Choices of technology could now involve up to five parastatal mills.

The original TEXCO proposal called for expansion by replicating the technology in each mill. The result would have been the purchase of capital-intensive equipment for some mills, labor-intensive equipment for others. This proposal was made and approved in spite of the national goals of using low-cost, labor-intensive, standardized technology. Neither the parastatal managers nor the cabinet-level decision makers related the decision to the national goals.

The eventual revision of the plans produced an even greater disparity between national goals and technology choice. The two mills finally chosen for expansion were those that had the highest social costs of production; moreover, they had relatively high costs at market prices. The more labor-intensive mills were not to be expanded.

Alternative models have been proposed to explain the choices of technology made by managers. One possibility is that the decision is a purely economic one; but the outcome in Tanzania seems not to reflect an effort to minimize costs at either market or shadow prices. Thus, the models of "economic man" seem inadequate to the task of explaining the decisions undertaken in Tanzanian textiles. The second chapter of this book suggested that such decisions may be made by an "engineering man," if price competition is insufficient to force managers to minimize costs. According to this model, managers might choose automated equipment to satisfy their desires for a modern, sophisticated plant, to reduce problems in managing labor, or to increase flexibility in adjusting output. The Tanzanian parastatal, however, had built both labor-intensive and capital-intensive mills. Its initial expansion plan was to increase capacity at each mill, using the technology in place. An engineering-man approach would presumably have consistently favored capital-intensive technology. To be sure, the eventual expansion decisions can be interpreted as a swing toward capital intensity, but the proposal from which the decisions were developed did not appear to intend such a swing. Neither the model of economic man nor the model of engineering man seems to provide an adequate explanation of the decisions actually made in Tanzanian parastatals.

PARASTATALS AND THEIR ENVIRONMENT

How is it that the Tanzanian parastatals—organizations that are a part of government—arrive at investment decisions that seem to be in conflict with state objectives? The explanation lies in the way in which these organizations respond to their environment, to other institutions in the environment, and to the problems they confront. To ensure their own survival and well-being, the parastatals have developed patterns of behavior that can produce outcomes inconsistent with national goals.

The National Development Corporation (NDC) was created as a holding company in 1965 in an attempt to fuse several roles played by development corporations. The central ministries oversaw the decision process of the NDC. The Ministry of Finance, wielding the ultimate power of the purse over both development and recurrent expenditures, allocated government funds of all types and granted final approval on all government or quasi-government expenditures. The other central ministry, for the period from 1965 to 1975, was the Ministry of Economic Affairs and Development Planning (Devplan). Together with Finance, Devplan wielded considerable powers of approval and allocation of the capital or development budget. It prepared the annual and five-year plans as well. Devplan was intended to be the guardian of economic policy and the controller of development.

In addition to the two central ministries, the Ministry of Commerce and Industries also had some influence on technology choice. This had been, essentially, a tutelary ministry concerned with licensing and registration of business activities, prices, and numerous regulatory activities. After 1968, however, it became the "parent ministry" for industry parastatals, and hence was more directly concerned with decisions in those institutions.

Other organizations that played an intermittent, but important, role in investment decisions were the banks, both Tanzanian and international; the international agencies such as the World Bank; and the various foreign institutions, both firms and governments, that were involved in particular decisions.

The evolution of the parastatal system in Tanzania may be divided into two periods. Between 1965 and 1970 the NDC expanded rapidly into a wide area of activities, guided in its efforts to develop the economy by the principles of the ruling party's 1967 Arusha Declaration, which set forth five themes on the development of Tanzania: self-reliance, rural development, equity, national economic control, and socialism. Although the organization had few personnel skilled in handling industrial development, there was a general sense of urgency to get things done quickly. Toward the end of the decade, a process of rationalization began. Numerous other parastatals were created, and the NDC contracted. Investment declined and projects moved more slowly to implementation. During this period, the NDC came under critical attack. Its investment proposals were subject to close scrutiny by the ministries, especially to the extent that the NDC needed ministry funds.

PARASTATAL INVESTMENT DECISIONS

Parastatals have participated in two major classes of investment. In some cases, the parastatal originated the investment procedures, selecting market, product, and manufacturing methods, and establishing, with or without collaborators, a firm to execute the particular productive activity. This class of investment will be designated as "original" investments. The second type of investment is one that does not result from any of the firm's own organizational processes, but from decisions made outside its organization. This class of "external" investments includes two subtypes. First the government acquired an interest in certain firms at the time of its Arusha Declaration. In addition, several new investments had been made, usually as a result of bilateral aid agreements, in which a parastatal was chosen by government as the executing agency for a project originally formulated without the involvement of the parastatal.

While externally originated investments cannot be regarded as direct outcomes of the organizational processes of the parastatals, they are nonetheless a product of the parastatal environment. Moreover, they become part of the portfolio of activities controlled by the parastatals and thus contribute to the pattern of outcomes that one observes as the technology "choices" made by parastatals.

Organizational Processes in Original Investments

The early period.   A senior official who was involved in much of the early high-level decision making in NDC reported that the evolution of decision procedures had been governed primarily by "a strong sense of urgency." He suggested that the main aim of the management team was to seek and to launch projects. The other stated aims of NDC became, in his view, the criteria used to decide whether the projects could actually be started. Of these other criteria, commercial viability was most emphasized. But the NDC was also required to attend to "the economic position and potentialities of Tanganyika as a whole."[2] This somewhat ambiguous objective was seen as implying, among other things, that projects should be directed toward the development and further processing of Tanzanian raw materials, and that undue risks should not be taken in search of profits. It was also open to numerous other interpretations, which could be used to argue for or against specific proposals.

The early sense of urgency is apparent in a speech made by President Nyerere to the board of the NDC: "We have reconstituted this board," he noted, "because of delay in getting decisions from the various ministries concerned; this is the reason why we have put this board together—consisting of Ministers."[3]

The effect of the underlying sense of urgency was to focus the attention of the early management on "packaging." Projects had to be "integrated"—a favorite term at the time—in such a way that speedy decisions could be made. But with little experience, few skilled people, and almost no reliable sources of information on technology, market, processes, and evaluation procedures, the NDC had to turn outside for the packaging of investment proposals.

The NDC sought information from the most readily available sources known to the management. More than three-quarters of the collaborating firms in the first three years of NDC projects were continental European, British, or East African firms. The predominance of continental European and British firms, which represented more than half the total, was probably due to the fact that the first general manager was an expatriate Frenchman, and several of the senior staff were also European or British. Later collaborators represented a somewhat wider range of nationalities. The standard operating procedure for obtaining information became, first, to place the problem before a foreign consultant as soon as a product market was identified, and, second, to select a consultant capable of dealing with the investment problem as a package. That is, the consultant should not have only technical or financial expertise, but should be equipped to deal effectively with all aspects of the problem.

Consultants were readily available. Multinational firms, especially those with interests in Africa, were themselves scanning for possible investments. Such firms were not only well equipped to provide the comprehensive assistance sought by NDC, but they were ready to act rapidly to avoid the loss of export markets or to preempt a potential new market. Other firms were eager to sell equipment, management skills, and other services, and had developed considerable expertise in putting together the kind of package required by NDC. Thus what began as a search for information almost invariably ended in a commitment to a collaborator.

Having established a relationship with a collaborator, the NDC passed the initiative for generating alternative solutions to the problem into the hands of the collaborating organization. A senior administrator, associated with the NDC from the government side during the early period, recalled that when questions of technical know-how were raised, NDC managers often replied that it was "technical know-who" that really counted. Because the know-how was all on one side, technological alternatives were appraised (if at all) largely on the basis of their appropriateness to the aims of the collaborating organization. This one-sidedness was not due to an inherently weak NDC position; indeed, the alacrity with which manufacturing firms were willing to turn themselves into instant consultants indicated considerable potential bargaining strength on NDC's part. But the need to proceed quickly to an acceptable package, coupled with lack of knowledge of alternatives, reduced both the will to insist on extensive search and appraisal and the awareness that such insistence could be of some benefit.

By the time a project reached the implementation stage, the collaborator was in virtually complete control, for two reasons. First, NDC had less expertise in this area than in any other. Second, NDC's scarce skills were needed for the start-up of other projects in earlier stages of development.

Thus, NDC's early procedures reflected its lack of skills and people, the diversity of its activities, and the perceived overriding pressure on the organization to start projects quickly. The need for quick results was met by dealing directly with manufacturing firms that were able and willing to deliver comprehensive solutions. Major alternative approaches were represented only by alternative potential collaborators, and as soon as the first acceptable package was discovered, the search would end. The evaluation of possible alternative technologies within the selected package was left to the collaborator; NDC rarely initiated technically oriented modifications.

One of the early investments originated from within NDC was Mwanza Textile Limited, incorporated in May 1966. The project was based on three major agreements. First, the NDC, the Victoria Federation of Cooperative Unions Ltd. (later Nyanza Cooperative Union Ltd.), and Amenital Holding Trust of Liechtenstein entered into an investment agreement to raise a sum of Sh 80 million to finance the facility.

An engineering consultant agreement was also signed by Mwanza Textile Ltd., Sodefra and Textilconsult of Vaduz, and Maurer Textiles S. A. of Geneva. The three foreign firms were required under the agreement to act as engineering and industrial consultants to Mwanza Textiles in designing and constructing the mill, installing machinery, and commissioning the mill into production. Remuneration for these services consisted of a fixed fee for the engineering work plus a percentage of the total cost of the mill, with a fixed upper limit for the total fee.

Finally, Mwanza Textiles Ltd. signed a management consultancy agreement with Textilconsult and Maurer. For management of the mill, a fixed percentage of profit before tax and depreciation was to be paid.

Sodefra, Textilconsult, and Maurer were all components of the Amenital Holding Trust. The same managing agents were partners in the Nigerian Textile Mills and were also involved in a textile mill in Ghana at about the same time. Amenital also had an interest in the Société-Alsacienne des Construction Mécaniques de Mulhause (SACM). Much of the equipment in Mwanza Textile Mill was SACM machinery delivered under a supplier's credit arrangement (involving a loan of Sh 60 million) that required repayment over eight years, with three years grace, and at an interest of 5.7 percent per annum. At least one observer felt this arrangement suggested that "the profits of Sodefra must have accrued through over-invoicing of the machinery."[4] The bankers used in negotiating the supplier's credit were French.

In 1968, the Tanzania Fertilizer Company was incorporated. This investment was announced by NDC as follows: "Kloeckner Industrie Anlagen of Duisberg, West Germany, is to set up a fertilizer factory for the company (NDC) on a turnkey basis at Tanga at a fixed price of Sh 118 million."[5] The agreement with Kloeckner was the culmination of a process that had begun with a 1967 market study by NDC. The study concluded that total consumption of fertilizer, then 30,000 tons per year, would grow to 80,000 tons by 1971, the earliest date at which it was thought a factory could be operating. The study was circulated to several leading chemical companies. One firm responded—Kloeckner-Humbolt-Dietz, the third largest company in West Germany.[6] The agreement stated that Kloeckner would select the most modern processes corresponding with the latest technical development in the chemical industry, taking into consideration the objective conditions prevailing in Tanzania.[7]

Several other, smaller projects were originated during NDC's early period. A contract was signed in 1968 with Danieli and Company of Italy for the supply and construction of a steel-rolling mill. The relationship with the Italian company originated in a proposal by NDC, which was circulated internationally. The Italian firm supplied plant and equipment at a cost of Sh 11.1 million and provided loans of about Sh 7 million, while taking a 20 percent equity position. Danieli also provided technical and managerial services.

Another enterprise, Tanzania Tanneries, was incorporated in 1968. A Swedish firm, Messrs. Ehrnberg and Sons Laderfabrik, took a minority position in the firm, supplied the machinery, and guaranteed to purchase 50 percent of the product in the form of wet blue semiprocessed leather, for further processing in its own plant. Earlier, in 1966, a factory for producing sisal bags was erected; Soc. Coll. Adriano Gardella, SPA, of Genoa, was the technical consultant to the project and took a minority position (20 percent of equity) in the firm, supplied equipment, and controlled the supply of spare parts.

In 1969, a somewhat larger project was established in partnership with General Tire International. General Tire took a 26 percent equity position on the basis of the supply and erection of equipment. It also arranged a loan and held a management agreement.

The investments of the early period generally followed much the same pattern: A European partner provided technical management and financial services, selected and supplied the technology, and implemented the chosen process.

Since NDC delegated the choice of technique to another organization, the actual decision must be seen as an outcome of the organizational process in the implementing firms. Further, since the chosen organizations were numerous, had a variety of objectives and of procedures for achieving them, and were all non-Tanzanian, it is not likely that any consistent Tanzanian standard of appropriateness would be met by all the projects. Thus, any simple economic model would probably be incapable of explaining how technologies were chosen. In the early period, except for textiles, none of the industries mentioned made more than a single investment. For textiles, one of the investments was "original," the other "external." Nevertheless, there is enough evidence to suggest something about the variations in the type of technology choice made and the possible alternatives foregone. The important questions are: Were there any alternatives? Is there any evidence that they were considered and rejected? What seem to have been the major factors determining the choices made?

Alternatives to the selected fertilizer process were available for consideration.[8] There are known anhydrite deposits in Tanzania providing the basic raw materials for gypsum, cement, or sulphuric acid. Ammonia (a by-product) from the oil refinery could have been combined with sulphuric acid to manufacture ammonium sulphate. In addition, several phosphate deposits could have provided raw material. Coulson suggests that these alternatives and others were rejected on the grounds that the small size of the market precluded an efficient scale of activity. This appears to have been a conclusion drawn by NDC.

The project that evolved was designed to make sulphuric acid from sulphur. The sulphuric acid is combined with phosphate rock to produce super phosphates and with liquid ammonia to make ammonium sulphate. The agreement reached with the collaborator was that sulphur, phosphate rock, ammonia, nitric acid, and potassium salts (the latter two for compounding into nitrogen-phosphate-potassium fertilizers) would be imported. Coulson produces evidence to suggest that the market study on which the project was based was inflated and largely "impressionistic," thus casting some doubt on the validity of the earlier decisions based on market size.

There is no evidence that the organization that chose the fertilizer process seriously considered alternative techniques. Indeed, it is unlikely that this firm could have taken a position as sole collaborator and then installed techniques suitable for exploiting local materials. Kloeckner is a manufacturer of fertilizers and fertilizer factory equipment, not a mining firm. The consortium would have had to include a collaborator with at least some experience in this field. On balance, it would appear that the salient fact about the technology chosen was that it was the kind with which the sole collaborator was most familiar. Perhaps it offered the highest financial returns to that firm.

The steel-rolling mill installed by Danieli is capable of rolling light sections from preheated billets. In some respects, the mill appears to have been designed with Tanzanian conditions in mind. It is not highly automated. After each pass, the sections are manually turned for rerunning. Other handling stages that could have been automated were not. On the other hand, there is some evidence that the mill was a leftover from a production line that had included continuous billet casting. The maximum billet size that can be run is 80 × 80mm, a nonstandard size that is often difficult to purchase on the world market. Although deposits of iron ore and coal have been known for many years in Tanzania, the possibility of early local billet production was quite remote in 1968. But the technology chosen was not entirely appropriate for the use of imported billet.

The equipment installed at the tannery by the Swedish firm was secondhand. This may be regarded as an appropriate choice, although there is no evidence that alternatives were considered. Ultimately, the equipment proved to be very troublesome. The choice can be most reasonably explained in terms of the product-cycle model.[9] Ehrnberg, a leather-goods manufacturer, discovered that it was costing them more to import unfinished hide and convert it to the wet blue stage (the early, simple, labor-intensive segment of tanning) than it would to import semiprocessed leather. Their interest was to secure a steady supply of wet blue hide. Tanzania was a major source of raw hides, was interested in processing its raw materials, and had relatively low labor costs. Ehrnberg became the collaborator in setting up a tannery that would produce finished leather but that would export half its production to Ehrnberg at the wet blue stage. The secondhand equipment was available because the Swedish factory was to replace its finishing equipment as the company shifted to importing wet blue. The breakdowns on the equipment did not seriously affect the production of wet blue, only of finished leather.[10]

The Tanzania Bag Factory was equipped with machinery procured and supplied by a Mr. Gardella, an entrepreneur who later became somewhat notorious in Tanzania as a result of the manipulations he performed on the Bag Factory and a Kenaf factory. The machinery supplied seemed to be of dubious quality and subject to constant breakdown; much revenue was collected by Gardella on spares that were obtainable only through his firms. Alternative sources of supply apparently were not considered.

The tire plant supplied by General Tire also used secondhand equipment. The machines operated efficiently and, in view of the narrow range of alternatives in tire-building equipment, probably represented the best choice for the conditions.

To summarize, the choice of technology in NDC projects in the period up to 1970 showed wide variations. Some of the equipment was relatively capital intensive, some relatively labor intensive. Some was mechanically reliable, some less so. All decisions were made by the collaborators and were more consistent with their objectives than with any other decision criterion.

The other organizational actors had very little influence over technology choice during the early period. The government planning and financial bodies generally took a relatively passive attitude toward NDC investments, for two reasons. First, for most of the period, NDC's board included the ministers for finance, planning (chairman), and industries. NDC's decisions, once past the board, were hardly likely to encounter serious objections from the ministries. Second, NDC was putting together financial packages that did not individually require large allocations of government funds; they caused no serious scarcity of funds. Thus, the planning system, based largely on budget management, did not seriously scrutinize the projects for purposes of controlling expenditures.

The later period.   The changes in the environment toward the end of the 1960s slowed the expansion of NDC and of the newly created parastatals managing NDC's former holdings and led to some modification of procedures and methods of project formulation. While staffing had improved somewhat in terms of quality and quantity, the parastatals still felt, with considerable justification, that they were not capable of independently evaluating all projects. Now, however, parastatal officials said they tended to seek a consultant who was *not* a manufacturer, machinery supplier, or potential collaborator but was, in fact, an advisory consultant. A package was still the ultimate aim, but the consultant was seen as a medium for exploring the alternatives open to the parastatal in formulating the project.

Since 1970, the initial studies of most original investment or expansion projects in manufacturing (including textiles) have been conducted by consultants who apparently had no interest in machinery or material supply for the potential project. Several, however, have had obvious interest in management contracts. While this interest may bias the choice of technology in some cases, it is likely to reduce the probability of installing inadequate equipment.

In this period, parastatals also changed the way they handled later steps in the process of investment project formulation. The consultant's feasibility study, once completed, became the document on which all subsequent steps were based. Subsequent steps tended to vary depending on the particular project. For small investments, the parastatal might try to put the project together with the assistance of the consultant and with financing supplied by the parastatal and the government. Equipment—recommended by the consultant—would be obtained on an international tender basis. On larger, more complex projects, financing became the dominant concern after the consultant's report. Potential financial contributors might require a second study or a revision of the first study; they might impose conditions on the sources from which equipment was to be chosen or on the type of equipment to be selected (e.g., secondhand equipment might be ruled out). The choice of technical, construction, and management advisors and contractors became subject to the approval of the financing agents.

The managing director of NDC cited a proposed pulp-and-paper plant as an example of the effectiveness of the new procedures. The idea of processing plantation softwoods into paper was extremely attractive to Tanzania. The venture would combine the processing of local raw materials with a potential increase in foreign exchange savings and earnings; moreover, the location of the mill would provide employment in an area of rural underemployment and initiate industrial development well away from the capital. As the managing director put it, in discussions, "The idea, at the level of national objectives, was so attractive that they wanted to do it next week."

Despite the pressure, NDC was able to follow the new procedures. Deeply afraid that going directly to manufacturers would mean an outcome like the fertilizer factory, the managing director retained an independent consultant from Sweden. At the same time, he made an attempt to obtain firsthand knowledge of the range of production possibilities and technologies by visiting several European countries and India to inspect other plants and machinery manufacturers.

After receiving the consultant's report, NDC asked for two outside opinions on its content. One opinion was offered by a Swedish state pulp-and-paper manufacturer; the second opinion was given by the World Bank. According to the managing director, these consultants enabled NDC to correct a considerable amount of overdesign and excessive automation in the proposal. Although implementation had not commenced at the time of this study, it seemed certain that the equipment would be put to international tender.

While decision making remained predominantly a function of organizational processes outside the parastatal itself, the procedures followed during the late period generally broadened the consideration of alternatives at some stages of project development—but not always, as the decision to expand capacity in the textile industry illustrates. TEXCO, the new corporation that took over the assets and liabilities of the textile firms[11] from the NDC group in 1975, was set up with an organizational structure almost identical to that of NDC. The senior staff members, from the managing director down, were almost all transferees from NDC. The departments established in the new holding corporation were identical in name and function to those of NDC.[12] The standard operating procedures and general approach to decision making in TEXCO were, for all practical purposes, identical to those at NDC at the time of TEXCO's establishment.

Pressure to expand the textile industry had been rising for some time, reflecting a wish both to reduce imports of fabrics into Tanzania and to process more of the country's cotton.[13] There was, on either basis, considerable scope for expansion of the industry, and one of the new corporation's first acts was to present an expansion program for the industry.

The initial submission to the economic committee of the cabinet, which proposed to replicate the capacity of all existing mills,[14] included a plan to establish five village weaving units. Each unit was to have twenty power looms. The village weaving sheds would receive yarn from one of the large spinning mills and return the gray cloth to the mill for finishing. The inclusion of the village weaving proposal was a response to pressure from some government leaders to increase the proportion of small-scale industrial activity. The scheme was not very detailed and was deleted from later versions of the expansion proposal.

In the revised proposal prepared by TEXCO,[15] the doubling of capacity in the Chinese-built Urafiki Textile Mill was reduced to a smaller balancing operation. This change was due, according to TEXCO, to the agreement between the two governments that Chinese aid funds would not be used at this time for textiles but on another project.[16] This did not preclude the purchase of Chinese equipment and services as an alternative; TEXCO investigated this possibility later, but was somewhat discouraged by the lengthy delivery quotations received from the Chinese. Most of the other proposed expansions remained the same, although some effort had been made to order priorities, since it seemed likely that insufficient equity funds would be available.

The revised proposal formed the basis for the TEXCO request for funds from the 1974–75 development budget. The total TEXCO request for equity funding was about Sh 200 million for 1974–75. The tentative allocation of development funds for the textile sector in that period was about Sh 20 million. Clearly, TEXCO was correct in presuming that some ordering of priorities might be necessary.

The strategy of the budget officers in allocating the equity funds was to try to realize the maximal increase in productive capacity for the smallest allocation of government funds in the shortest time. They lacked detailed information, however, and moreover were attempting to allocate some Sh 2.4 billion during a series of budget hearings with different agencies over a period of two or three weeks. As a result, funds tended to be allocated to those projects that required the lowest allocation *in that year* relative to the productive capacity planned and that seemed to be in the most advanced stage of preparedness. The bias was toward big projects with low initial equity expenditures. In other words, the organizational processes of the budget mechanism encouraged the making of small starts on big projects, if equity funding was required, and the proposal of projects in which equity funding was not required or was minimal. In this situation, TEXCO chose to push for the expansion of the largest mill other than Urafiki, which was Mwanza Textiles, and the mill for which an expansion program had already been prepared by the management team and the private partners, Sunguratex.

Members of the Industrial Division of Devplan pointed out that the choice of the Mwanza and Sunguratex Mills implied a choice of capital-intensive technology. But both government and parastatal officers were conscious that the possibility of further Chinese aid, in the form of a labor-intensive mill like Urafiki, was ruled out at a high level. Neither group was enthusiastic about trying to obtain Chinese equipment on commercial terms. The managers of TEXCO were afraid of delay, and they had delivery quotations of as long as two years to support their fears. The planners, without sufficient data on the advantages of Chinese equipment, were unwilling to insist on a course of action that meant delay. Both groups tended to feel that the subject was closed.

The proposal to expand the two capital-intensive mills was accepted, with a proviso that the World Bank, which was to send an industrial-sector mission to Tanzania shortly afterward (in October 1974), should be asked to include a textile specialist in the team. The purpose of this request was to obtain the Bank's view of the expansion proposals and to encourage some tentatively expressed interest in financial support of the Mwanza expansion.

At this point, then, the operating procedures of both the parastatal (supported by its parent ministry) and the central ministries controlling the budget had tipped the scales heavily in favor of choosing the more capital-intensive, higher-cost techniques in the textile expansion. True, the actual equipment had not been chosen; but once a specific mill had been designated for expansion, there are compelling reasons to choose technologies closely related to the existing equipment.

The World Bank mission conducted its analysis of the industrial sector in October 1974.[17] The group included a textile team consisting of a textile analyst, who visited all the major mills to assess the potential for expansion of the industry; and a financial analyst, who investigated the financial implications of an expansion program with the TEXCO Division of Planning and Finance and who visited one mill, Mwanza Textiles.[18]

The World Bank mission endorsed the provisional decision to expand the two mills chosen by TEXCO and announced that it would recommend that the Bank finance the Mwanza expansion. Asked why it endorsed the choice of Mwanza, when Urafiki and Kiltex had been shown to be operating at lower cost, the mission pointed out that World Bank standard operating procedures would not permit a project financed by the Bank to be equipped with machinery from a nonmember nation, which China was. Since the Urafiki Textile Mill was equipped with Chinese machinery, a replication of that facility could not be recommended.

The range of technology to be considered narrowed under the influence of successively added layers of organizational procedures. TEXCO's original expansion proposal, submitted to the Economic Committee of the Cabinet, included a range of technologies that went beyond the range eventually selected. In addition to proposing expansion of all mills, the proposal included the village weaving units. The second version of the expansion proposal, compiled in the light of TEXCO's own reassessment of its priorities and the possibility of implementing the proposal, dropped the village weaving proposal.[19] As a result of the interaction between TEXCO and Devplan, the two most labor-intensive mills were removed from the immediate program, leaving a provisional choice of Mwanza and Sunguratex—the two most capital-intensive mills. Since the World Bank procedures did not allow for Chinese equipment and no search for other suppliers of such equipment was made, the large-scale, labor-intensive option was eliminated.

One cannot conclude that a consistent bias against labor-intensive methods caused a choice of capital-intensive technology. A slight rearrangement of circumstances (e.g., a different result in negotiations with the Chinese or a shift in allocation emphasis in Washington) could have changed the outcome significantly. Nor were factor price distortions responsible for the capital-intensive choice. At current prices, the least-cost mills were the ones deleted.

The most plausible conclusion is that, in the Tanzanian context, technology is not selected for its contribution to the operation of the firm; rather, technology choice is an outcome of organizational processes and interactions. The organizations involved each have an array of goals. Progress toward implementation of the common goal represented by the project under consideration gives rise to problems and choices. Each organization solves its problems and makes choices in the light of its own objectives and using its own procedures. Incompatibility between the common goal and individual organizational goals and procedures may be accommodated in several ways. The organization may retire from the interaction, or the definition of the common goal may be changed. A new organization may enter. At some point, a satisficing equilibrium is achieved. Each organization perceives the state of the common goal as adequately compatible with its objectives. The interaction ceases. The project is defined. At some point during the process, technology is selected as an outcome of the organizational interaction.

The investment procedures developed by the parastatals in the late period seemed likely to ensure that projects were better prepared and equipment more carefully chosen than in the earlier period. The managing directors of both NDC and TEXCO made it clear that their organizations were very sensitive to the criticisms leveled at earlier projects. The involvement of independent consultants, the invitation of second opinions from the World Bank and other prestigious international institutions, and the insistence on international tendering were all organizational responses calculated to prevent the repetition of past mistakes. At least, as one managing director pointed out, if a "bad" project is implemented under these conditions, the parastatal is "in the best of company."

Nevertheless, even where the parastatal perceived some freedom of technique choice, the new procedures were not directed at ensuring that Tanzanian criteria govern that decision. In the pulp mill case, the general manager admitted that, while he was personally concerned about the choice of technique, his primary concern was that the project not be equipped with mechanically inadequate machinery. He emphasized that the care with which NDC had been planning the implementation of the pulp project was due mainly to the project's size and complexity. Had the project been less complex and hence easier to finance quickly, he might not have been able to withstand the pressure to act quickly and would probably have had to deal directly with a collaborator. As it was, he claimed, the Ministry of Industries had been sharply critical about the expenses for consultants that yielded no visible development of the project.

The organizational processes of other institutions continued to constrain the choice of technique. For example, the External Finance Division of Treasury negotiated with foreign donors and lender governments to determine the allocation of foreign aid and loans among the projects in the Development Budget. The donors typically differed in their preferences as to the areas in which their aid or loans should be used and in their schedules of terms and conditions for use of the funds. Neither the Treasury officials nor the officials of the foreign governments were concerned with the choice of technology in a particular project. Consequently, they were often oblivious to the impact of their bargaining outcomes on individual projects. For example, as the NDC director of planning and finance explained, NDC might have chosen Italian technical advisors for a proposed project while Treasury had arrived at an agreement that Dutch aid would be used for the project equipment. In this situation, the recommendations of the technical advisors might have to be changed to accommodate Dutch procurement rules, or a new financing source might have to be arranged. If the new source were to be found by NDC, it might have to be a foreign equipment supplier offering credit. The ultimate choice of equipment thus was an outcome of the organizational interaction.

The NDC official cited several cases in which the organizational processes of Treasury and of the aid donors had affected equipment choice. In one case, a bicycle plant, external financing was switched from Dutch to German to Swedish sources over a few months. The Dutch and German procurement rules required preference to Dutch or German suppliers but permitted procurement from other countries (including certain developing countries) if cheaper delivery of equivalent equipment could be proved. Sweden put few restrictions on procurement and did not exclude any sources. The progression of changes in the bicycle case was toward more freedom of choice, but it could just as easily have been in the opposite direction.[20]

Organizational Processes in External Investment

A large fraction of the investments made by Tanzanian parastatals (e.g., about half the current investment program in textiles) is projects that originated outside the parastatal system. In these projects, international political decision making is directly involved in some or all of the processes from origination of the project to its implementation. Typically, during some high-level political interchange between governments, aid is requested or offered, and promised or accepted, in the form of an industrial project. Such arrangements vary in the degree to which the project is initially specified. Sometimes there is agreement on general assistance in some area of industrial activity. Other projects may be completely specified: a complete factory for the production of a precisely defined product mix. The arrangements also vary in the degree to which they fit into the plans of parastatals concerned with the relevant sector. Sometimes the arrangements in effect provide a solution to a problem already being worked on. At other times, the parastatal may be presented with a project that it has never contemplated.

One of the earliest examples of an externally originated investment was the Urafiki Textile Mill. This project originated during high-level contacts between the Tanzanian and Chinese governments before 1965.[21] In January 1965, an agreement was signed between the two governments for the establishment of a textile mill in Dar es Salaam. The total capital cost of the mill was to be Sh 50 million, to be provided as an interest-free long-term loan to the Tanzanian government from the People's Republic of China. The project was planned as two integrated spinning and weaving units with 20,000 spindles each, 6,000 and 378 looms for the production of finer and heavier fabrics respectively, and a finishing unit.

Chinese aid projects had some special characteristics that bore on the choice of technology. Once committed to a Tanzanian project, the Chinese tended to move with unobtrusive but nonetheless concentrated swiftness toward completion. Very often, little was known about the project in Tanzania outside the circle of Tanzanians directly involved in implementation. Dealings on the project remained at a consistently high government level. The civil service often had virtually no information about the project. Relatively large numbers of Chinese technicians, managers, and workers would be employed on the project but rarely would be seen away from the site.

The rapid, unobtrusive, almost secretive, implementation process produced at least two results unique to Chinese projects. First, there was little discussion of the project by planners and other government officials, and consequently few objections would be raised on any aspect of the project. Second, the air of secrecy and isolation surrounding the project sometimes lingered after the Chinese handed over the plant, making it somewhat difficult to assess the nature of decision making during implementation. A further feature of Chinese projects was a relatively rapid transfer of technology. Tanzanian management and staff were generally able to operate and control the institution entirely very shortly after its completion.

The textile plant was designed in detail in China without an examination of Tanzanian conditions. No NDC staff member interviewed could recall any consultation on technology or any modification to the equipment design resulting from Tanzanian requirements or suggestions. One senior official remarked that the project was so completely set apart from the Tanzanian economy during its development that the first production trial runs produced unsalable fabric—not faulty fabric, but simply not the kind of fabric Tanzanians buy. The original output was, he alleged, still in stock "somewhere." The fabric design was quickly adjusted and the mill operated satisfactorily thereafter.

A second Chinese investment went into production in 1968, with installed capacity of 1,000 tons per annum of hoes (*jembes*), machetes (*pangas*), axes, ploughs, and plough parts. Total cost of the farm-implements project was Sh 3.48 million, financed entirely by interest-free loans from the Chinese government. About twenty-six Chinese technicians and over 300 Tanzanians formed the construction team. The pattern of technology choice and plant design was the same in this case as for Urafiki textiles.[22]

A more recent investment provides an example of how the government may plug a politically agreed-upon project into a design already conceived by a parastatal. In 1972, NDC was planning expansion of its capacity in chrome tanning. In addition, the corporation planned to integrate the production of leather, leather goods, and leather shoes by building a complex of several factories around a large tannery at Morogoro. A second tannery, without the downstream industries, was planned for Mwanza. Based on its somewhat mixed experience with the Swedish-built tannery, NDC appeared to be applying a careful search for suitable technology and collaborators. An NDC mission visited some dozen tanneries in various developing countries, seeking to avoid errors in technology choice and paying particular attention to materials-handling methods. In addition, in line with later operating procedures, an independent consultant had been retained for market and product advice. But by the time the NDC study group returned to Tanzania, an agreement had already been signed, during a state visit to Bulgaria by senior members of government, specifying that the Morogoro tannery would be supplied by the Bulgarian state firm of Techno-Export.[23]

The initial reaction of NDC management was to protest, through the parent ministry (Industries), that the process of selecting machinery supplies was not complete. Moreover, the survey team had visited a tannery in Algeria that had been having problems with equipment supplied by the Bulgarian firm. The NDC planners wanted some assurance that the problems experienced in Algeria would be resolved. It appeared, however, that the political decision was firm. In the comprehensive feasibility study of the Morogoro complex prepared by NDC's consultants, the choice of equipment suppliers for the tannery was described as follows:

> ... the proposed tannery aims at turning out high-quality chrome tanned leather, for which it is proposed to install modern machinery equipment for leather tanning and finishing.
>
> In this connection, National Development Corporation (NDC) contacted several machinery suppliers, and finally a contract was signed with Bulgarians for the purchase of complete plant from them. ...[24]

The consultants' report added that the Bulgarian firm would "themselves acquire some of the machines from Czechoslovakia, France, and other places to supplement the machinery which they can manufacture." Thus, the choice of technology was passed to the collaborators by virtue of the high-level government agreement.

In 1976, TEXCO was involved in establishing a garment factory in Dar es Salaam. During the discussions held to decide on development budget allocations for the third five-year plan, the strategy of developing the garment industry was discussed at some length. Devplan felt that the industry was ideally suited to contribute to the government's policy of dispersing industry. The manufacture of garments is relatively simple and can be conducted efficiently at almost any scale of operations. Hence, the planners argued, the strategy should be to

develop a number of small-scale garment manufacturing establishments in various parts of the country. Moreover, because garments
could be manufactured at a village-industry level, the production could
be integrated with the government's village development program,
which was already well under way. While the management of TEXCO
was somewhat skeptical of the proposal, it agreed to examine the
possibilities.

Before any detailed plans had been formulated, however, an
agreement was signed between the Romanian and the Tanzanian
governments for the supply of a garment factory under Romanian
technical assistance. The Romanian proposal involved setting up a
new factory in Ubungo, Dar es Salaam, as a fully owned subsidiary of
TEXCO. The total cost of the project was about Sh 31 million.
Equipment would be supplied by the Romanians under supplier's
credit terms. The factory would employ about 800 people, of whom a
small number would be Romanian technical assistance personnel.
Annual production would be 1,500,000 pieces of men's pants, wom-
en's dresses and slacks, and children's dresses.

Once again, a high-level visit to a potential donor country had
transformed variables and alternatives into fixed parameters. In this
case, the choice of technology, particularly the choice of scale and
location, was set by the intergovernment agreement.

The technology choices resulting from politically initiated investments are as varied as those of "original" investments. The technology
built into the Urafiki Textile Mill is relatively labor intensive and has
proved to be relatively efficient. The mill has been the most profitable
of the textile firms operated by TEXCO. The choice of labor-intensive
equipment by the Chinese donors appears to have been deliberate.
According to a former NDC general manager, the Chinese were
producing a range of textile equipment of which the type installed at
Urafiki was the most labor intensive. Similarly, the farm implements
plant appears to have been well equipped with a blend of manual and
semiautomatic equipment. There were some doubts about the design
of the production line, which made it necessary to restrict products in
that section of the factory to the production of one item at a time.

The other two investments, the tannery and the garment factory,
were somewhat less satisfactory in terms of technology choice. The
tannery became a source of some contention between NDC and

Techno-Export. NDC management felt that the tannery was going to be unnecessarily capital intensive, particularly since a very expensive building[25] would be required to accommodate the materials-handling methods (using overhead cranes) specified by the Bulgarians. NDC claims that it would have opted (as it has in other tanneries) for floor handling methods requiring more labor, cheaper equipment, and normal buildings. In this case, the parastatal itself viewed the technology as inappropriate.

The design of the garment factory seems to conflict with the general goal of developing small-scale, rurally located industry. Instead of employing 800 people in one factory in the capital city, it would have been possible to employ eighty people in each of ten small plants in different towns, with little loss of efficiency.

Like the parastatals' original investment, the externally originated projects show a wide range of technology choices—in large part because these decisions are a function of the operating procedures of the donor organizations. This outcome does not appear to be inevitable. There are some indications that parastatals may be both willing and able to achieve more influence over technology choice. But there is also evidence of forces that will continue to constrain deeper involvement by the parastatals.

A politically originated investment moves through two major phases: first, the process of contact between the governments; second, the process of implementation, which takes place at a lower level, usually between organizations controlled by the respective governments. There is probably very little that can be changed about the process of origination. Leaders of poor, developing countries will continue to seek and to be offered aid projects. There is no evidence that any of the Tanzanian political projects were in themselves inappropriate; all were in industries that fit quite obviously into the apparent program of industrialization.

At the point of initial agreement, however, the understanding of the final nature of the project is usually relatively vague. The government officials or political leaders agree on "mutual assistance" or a "project to aid the development of the textile sector" or some similar, general notion. In no case examined did the initial decision include the specifics of technology. In all cases the final project was defined on the basis of a study carried out later, usually by the donors. Hence, in theory, there is some period of time during which formulation of the project is open to discussion and Tanzanian views on technology could be introduced into the process. It is primarily the desire of both parties for quick implementation that limits this influence.

CONCLUSION

In the "original" investment projects, the parastatals appear to be in a position to make choices of technique. These choices may be seen as outcomes of the interaction of organizational processes of the involved institutions. During the early period of parastatal development, the demands of their environment forced managers to emphasize speed of development. The inherent weaknesses of the parastatals led them to develop procedures that permitted speedy implementation of projects through packaging and through assignment of technological decisions to collaborators who could deliver a complete project. The outcomes of the early organizational processes varied widely. Some equipment was relatively capital intensive, other equipment was relatively labor intensive; some was new, some was used; some equipment was mechanically efficient, other equipment less so. Since all equipment was chosen by the collaborators, conformity to any consistent Tanzanian criteria was a matter of chance.

In the later period, parastatals responded to criticism of the earlier projects by modifying their procedures. The new processes appear to have provided some safeguards against mechanical inefficiency and supplier dishonesty. However, since some sections of government were still pressing for rapid development, the newer procedures were not always followed. Moreover, because the parastatals were more reliant on the development budget than in the past, the organizational process of managing foreign aid and loans became more important, as did the processes of such institutions as the World Bank. Consequently, the choice of technology was still an outcome of the interaction of the organizational processes of institutions, most of which had little direct interest in applying Tanzanian criteria to the equipment needs of particular projects.

For the parastatals themselves, production appeared to be the dominant goal. Costs of production were relatively unimportant, since prices were adjusted to cover costs in the state-controlled environment. Timing or finance issues were more important for managers.

When industrial projects in Tanzania were originated outside the parastatal system, parastatal management had limited freedom of choice of technology. These investments resembled the original investments of the early parastatal period in that the bulk of decision making was passed to a single external institution at a very early stage of implementation. In general, the decision of the donor appeared to dominate. The variations in technology chosen in these externally originated investments are as wide as those observed in the investments originated by the parastatals themselves.

Part 2

Channels for Transferring Technology

8

Channels for Technology Transfer: The Petrochemical Industry

ROBERT STOBAUGH

Whatever the kinds of technologies to be transferred, the firm faces alternative means for effecting the transfer. The choice is of considerable importance to governments as well as to the enterprises involved in the transfer. This chapter highlights the major factors that determine the choice of channels used to transfer technology in the world petrochemical industry. Two questions are examined in detail: How does the technology owner decide whether to invest in a facility to use the technology or to sell the technology for use by an unrelated party? And, if the decision is to invest, how does he or she decide whether the facility should be wholly owned or a joint venture with an unrelated firm?

Petrochemicals are chemical products made from petroleum or natural gas. The petrochemical industry starts with feedstocks obtained from oil refineries and natural gas processing plants. It then transforms these feedstocks through a number of process steps into petrochemical products: bulk chemicals, polymers, fibers, rubbers, solvents, and bulk fertilizers. From these petrochemical products are made final products for consumers—garden hoses, steering wheels, clothes, tires, and complex fertilizers, to name just a few of thousands.

Most manufacturers of petrochemicals are large international firms (either oil or chemical), and most petrochemicals were originally commercialized in a few industrialized countries. By now, however, petrochemical plants have been established in many other countries. For the governments of these technology-importing nations, the selection of channels raises questions about the degree of dependence on a few firms from a few countries.

This chapter is based on a study of nine petrochemicals (referred to as the Nine Products) from the beginning of their product lives through 1974, at which time they ranged in age from twenty-five to sixty-seven years.[1] During these years, market growth in the originally innovating country and abroad caused a mushrooming of production facilities (see figure 8–1).

A total of 537 manufacturing units, each individually designed to make one of the Nine Products, were built worldwide (excluding a relative few in planned economies). This chapter uses data for 515 of these plants. Because some used technology from more than one source, the 515 plants involved 592 technology transfers, for 586 of which data were available. By definition, a "technology transfer" occurred any time a plant was built. In other words, technology transfers include cases in which a company used technology it had developed itself as well as cases in which one company transferred technology to another unrelated company.

Three types of firms were involved in technology transfers:

1. Firms that had initially commercialized a *product* (and in every case first exploited the technology in a company-owned facility). Although patents could theoretically protect the market position of the product innovator for some years,

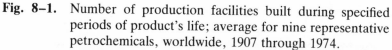

Fig. 8–1. Number of production facilities built during specified periods of product's life; average for nine representative petrochemicals, worldwide, 1907 through 1974.

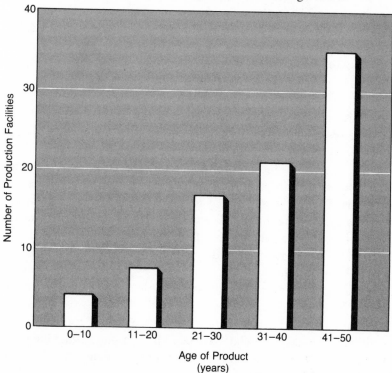

SOURCE: Industry trade journals plus author's questionnaire.
NOTE: The number of facilities existing each year between 1907 and 1974 was determined; facilities were then classified by age of product. Not all Nine Products are as old as 50 years, the life span covered by this exhibit. The average number of the Nine Products that existed in each decade was: 9 in first and second; 8.5 in third; 5.2 in fourth; and 3.2 in fifth. Thus, for example, an average of 8.5 products had entered the third decade of their life (age 21–30) as of 1974, and an average of 18 production facilities had been built during that decade of each product's life.

        many petrochemicals (including the Nine Products) had been discovered so many years (sometimes 100 or more) before commercialization that a product patent was not available to the commercializing firm.

2. Firms that had developed a new commercial *process* for manufacturing a product that was already being produced commercially by an existing process.

3. Firms that purchased the technology.

Oil refining or chemical manufacturing was one of the principal businesses of all of the product innovators and of most of the process innovators. Some process innovators were engineering contractors, which sold a package of design, purchasing, and construction services along with the technology. They rarely manufactured petrochemicals; indeed, of the 139 transfers of technology by an engineering contractor, 138 were arm's-length sales to an unrelated party. Bechtel, Fluor, Lummus, and Scientific Design are well-known examples of such firms.

Three different channels were used for the transfer of technology: the sale of technology to an unrelated firm, investment in facilities partially owned by the technology owner, and investment in facilities wholly owned by the technology owner. Any given firm at any one time could be employing all three channels simultaneously in different locations. The sale of technology to an unrelated party almost invariably involved the granting of a license allowing the purchaser to use the technology; thus, this channel is referred to as "licensing." Hence, the term "licensing" is not used to describe the transfer of technology to a facility owned, either wholly or partially, by the technology owner, even though such a transfer sometimes was also accompanied by a license.

Somewhat more than half of the 586 technology transfers were made by licensing (to unrelated parties). In the transfers that involved ownership, wholly owned subsidiaries heavily outnumbered the joint ventures, as shown in figure 8–2.

THE LICENSING VERSUS INVESTMENT DECISION

The benefits and costs to the technology owner—and hence the decision whether to license or to invest—were found to depend on several factors, most significantly on whether the transfer was international or domestic.[2] Other important factors included the characteristics of the country in which the facility was to be located, the competition faced by the technology-owning firm, and certain characteristics of the technology-owning firm itself.

To find the major determinants of the license-versus-investment decisions of firms that could potentially choose either action, the

Fig. 8–2. Number of transfers of technology via different channels; 586 transfers, worldwide; nine representative petrochemicals; through 1974.

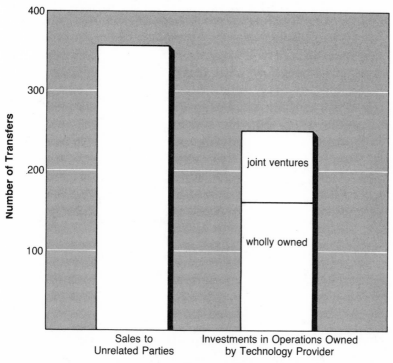

SOURCE: Industry trade journals and author's questionnaire.

transfers of technology by engineering contractors were excluded from the statistical analysis (see Appendix to this chapter). Thus, the results described here were based on 447 transfers by manufacturing firms.

International Versus Domestic Transfers

Seventy-six percent of the international transfers, but only 42 percent of the domestic transfers, involved licensing (i.e., sales to an unrelated party). Why does crossing a national boundary make a difference?

One important factor is that operating abroad requires more resources than operating in one's home country. Most petrochemical firms have less extensive operating experience in any single foreign country than in their home country. This translates into a lesser availability of resources—especially management skills and knowledge of local conditions—for foreign operations than for domestic operations. Thus a company wishing to transfer technology abroad tends to license a firm that already operates in the foreign country and already has management in the country as well as a knowledge of local conditions. Managers also perceive that there is greater risk in operating abroad than in one's own country. These risks range from misjudging the market to nationalization.[3]

Two other differences between international and domestic transfers were of less importance for the Nine Products. The first is the possible effect on competition. Most petrochemical firms have larger sales for any given product in their home market than in any single foreign market; thus, greater competition in their home market is likely to be more costly to them, in terms of lower prices and lost sales, than would greater competition in a foreign market. Even licensing an overseas producer may pose some threat to domestic business. Although the sale of know-how usually restricts the purchaser to manufacturing in one or more specified countries, it ordinarily cannot limit effectively the geographical territory in which the product is sold. By the time in the product life cycle at which know-how begins to be sold, any patent protection is likely to be on a process rather than the product. Still, because of tariff and freight charges and the lesser security of supply inherent in importing, a licensee manufacturing abroad is likely to be less of a competitor in the technology seller's home market than would a licensee manufacturing domestically. In any case, the desire to protect the domestic market does not seem to have been overwhelming in affecting the decisions on transferring technology. Of the initial sales of technology for each of the Nine Products, four were within the home country of the technology seller.

Another difference between international and domestic transfers is that some nations receiving technology prefer locally owned and controlled companies to foreign-owned ones; Japan, Mexico, and India are examples.[4] Governments want to keep as much control as they can in the hands of nationals.[5] Although the desire of some technology-importing nations for local ownership is important in individual instances, it does not play an important role in explaining the worldwide pattern of transfers used for the Nine Products. There were simply too few instances of transfers to countries with a strong policy favoring local ownership to make a significant difference in the totals.

Characteristics of the Country Where the Technology Was to be Used

Two characteristics of the nation where the technology was to be used affected the licensing vs. investment decision: the stage of industrial development and the size of the market.

There was a greater tendency to use licensing to transfer technology to developed countries than to less developed ones. Firms capable of purchasing the technology and operating the facilities are more likely to be found in developed countries. One could argue, of course, that the local entity needs nothing but money to purchase the technology, as a potential purchaser could hire someone to manage the project, operate the facility, and market the output. But this pattern has seldom been followed in the petrochemical industry, and most firms purchasing technology have had at least some knowledge of the industry.

Mixed evidence was found concerning the impact of the size of the host country's market on the decision to license or invest. Managers of technology-owning firms stated in interviews that they have a greater preference for investment (as compared to an arm's-length sale) when the technology is to be used in a large-market country, for the simple reason that they want to use available managerial resources in countries with the greatest economic potential.[6] But an analysis of the 447 transfers showed that, other things being equal, a company transferring technology into a large-market country was more likely to sell it to an unrelated firm than was the case in transfers to nations with small markets.

There are several possible explanations for this discrepancy between what managers said and what they actually did. First, it could be a statistical artifact, resulting from the fact that most large-market countries are also highly developed ones. (There was a statistically significant correlation between these two variables; see Appendix to this chapter.) In other words, the likelihood that there would be a local entity able to purchase the technology was measured by the market size variable as well as by the developed/less-developed country variable. Second, large markets may have been more competitive, decreasing the attractiveness of a foreign direct investment there. The third possibility is that licensing arrangements in large-market countries may have been part of reciprocal licensing agreements under which firms grant licenses to other innovative firms within the industry, in order to be allowed to purchase technologies owned by these other firms. Such "reciprocal licensing" can result either from formal agreements between firms or from informal understandings that reflect a live-and-let-live practice in the industry. It is most likely to occur between large firms, which are likely to be headquartered in large-market nations. There was no evidence, however, to support the notion that this type of licensing was widespread.[7]

Though all these factors may have contributed to the inconsistency between managers' expressed views of licensing in large-market countries and their actual behavior, no one explanation is compelling.

Competition Faced by the Technology-Owning Firm

Data for the Nine Products show the influence of competition, especially the presence of engineering contractors, on the decision to sell technology. A company manufacturing a product with its own technology does not typically make its first sale of technology until competition is well established. Considering the first licensings for each of the Nine Products, one finds that the median number of other companies with alternative technology for making the product was eleven (of which ten were using it in manufacturing). This initial sale occurred some twenty-two years (median) after the original commercialization, about the time when engineering contractors were beginning to appear on the scene with their proprietary technology. In fact, for four of the Nine Products, engineering companies sold the technology before any manufacturer did so.

Once the initial sale of technology was made for a product, licensing by manufacturing firms became widespread, accounting for 211 sales of technology as compared with 139 by engineering contractors. Engineering contractors seemed to be an important factor in influencing manufacturing firms to sell technology: the greater the number of engineering contractors owning the technology, the higher the probability that a manufacturing firm would license rather than invest in its own facility. This finding was true for both domestic and international transfers. An increase in the number of manufacturing firms owning technology also increased the likelihood of a sale of technology in a domestic transfer by a manufacturing firm, but it did not seem to affect international transfers. This result suggests that increased competition by manufacturers has a greater impact on a firm in its home market than it does in its foreign market.

There are several possible reasons for the influence of competition on the licensing vs. investment decision. A manufacturing company that owns technology and uses it to manufacture a product can face competition in two different markets: the technology market and the product market. When there is only one owner of technology and one manufacturer (and these two entities were initially the same for each of the Nine Products), the return on investment is typically highest. Monopoly profits accrue both to the monopoly on technology and to the monopoly on manufacture.

The monopoly profits erode when there is an increase in the number of companies owning the technology or in the number of manufacturers (or both). Thus, if the product innovator sold the technology to another company, the monopoly profits due to manufacturing would shrink because of the increased number of manufacturing competitors.[8] The technology seller would hope to recoup this loss in monopoly profits in the price of the technology.

But the minimum technology price acceptable to the seller is likely to be higher than the maximum price acceptable to the technology buyer, for a number of reasons.

Because of the significant scale economies involved in the manufacture of petrochemicals, the market is likely to be largely accounted for by a few manufacturers, each of which is likely to have an important share of the market. Each of these will therefore bear an important share of the losses of monopoly profits borne by the collective group of manufacturers. This potential loss of monopoly profits represents an important barrier to technology sales when there are just a few manufacturers; but when there are many manufacturers, this loss, for at least some manufacturers, will be very minor. Hence, the propensity to license rather than invest increases with an increase in the number of manufacturers.

Other factors also encourage firms to use their knowledge in their own facilities rather than sell it to unrelated parties. First, the seller of technology inevitably knows more about the technology than does the potential buyer, since the seller will not want to divulge all its secrets before the sale; because of its greater uncertainty, the potential buyer is willing to pay less than the seller knows the technology is worth. Second, profits can be maximized through discriminatory pricing of a kind that is feasible when the firm exploits its own technology in different markets, but that is much more difficult to accomplish when a number of unrelated entities are involved. Third, when there is only one seller of technology and one prospective buyer, the resulting bilateral concentration of power leads to an indeterminate or unstable bargaining situation, because typically there is a range of prices for the technology that would be satisfactory to both the seller and the buyer. Fourth, there is no futures market for knowledge—that is, no way to encourage investment now in order to generate knowledge for sale in the future—so the technology developer, in effect, creates its future market by using its knowledge itself. Fifth, when knowledge is transferred across national boundaries, it is easier to minimize total taxation by different national governments if the transfer is made within the firm rather than through an external market.

To offset these factors, the technology purchaser must have some advantage over the technology seller in exploiting the technology; otherwise, there would be no sale.

As competition in the technology market and the product market increases, however, all these factors become less significant.[9] The net result is a much greater tendency for the technology owner to license a prospective user.[10]

Resources of the Company Owning the Technology

The size of a firm, as measured by annual sales, played an important role in determining its propensity to sell technology. Compared with large manufacturing firms, small manufacturing firms were much more likely to transfer technology through sales rather than through investments in their own manufacturing facilities. There are several possible explanations for this pattern.

First, small firms are less able to bear the risk of the capital investment than are large firms. This is an especially important consideration in petrochemicals, where economies of scale in manufacturing dictate that a relatively large—and hence costly—plant must be built to be economically competitive. Since greater risk is involved in investing abroad than at home, the size of the firm had even more influence on the international transfer decision than on the domestic transfer decision. That is, small firms were even more likely to opt for licensing rather than making an investment when an international transfer was involved than in the case of a domestic transfer.[11]

Second, the marginal costs of the additional management needed to operate a new facility, especially one located abroad, are likely to be higher for small firms than for large firms. Since each new executive added to the payroll represents a commitment by the firm to pay salary, benefits, and expenses for a number of years, as well as a commitment on the part of existing management to train the new executive, the addition of executives for a new facility is likely to loom larger for a small firm than a big firm. Furthermore, a big firm—with its likelihood of greater foreign experience—is more likely than a small firm to have in its employ one or more foreign-experienced executives available for such an assignment.

Third, a small firm is likely to have a smaller market share in the product than a large firm (since, in order to lower risk through product diversification, the small firm must spread its resources over a number of products instead of concentrating on just a few). Therefore, the small firm would assign lower absolute cost to the creation of another manufacturing competitor, either in its home market or abroad, than would a large firm. For example, a firm with a 10 percent market share probably would experience a lower absolute cost because of competition from a new entrant than would a firm with a 20 percent share. (But in some cases, the firm with the larger market share would be better able to withstand the additional cost because of the higher profit margins sometimes associated with larger market shares.)

Given these general considerations, it is not surprising that engineering contractors almost always license rather than invest. They typically are smaller than manufacturing firms and have no experience in manufacturing petrochemicals—and thus no market share.

One final characteristic of firms that was tested for importance in the licensing vs. investing decision is national origin. About half of the total transfers of technology by manufacturing firms were made by U.S. firms (239 out of the 447). There was no significant difference between U.S. firms and non-U.S. firms in their propensities to sell technology, as opposed to using the technology in their own facilities. This was true both for the international and the domestic transfers of technology.

License or Invest: A Summary

In summary, when managers of manufacturing firms considered two ways of transferring technology, a sale of technology to an unrelated firm and an investment in a production facility, the propensity to sell the technology was greater:

- When the transfer was an international one;
- When the technology was transferred to a developed, large-market country;
- When a relatively large number of companies owned technology to make the product in question; and
- When a relatively small company was making the decision.

Whether the company was a U.S. firm or a non-U.S. firm did not seem to affect the decision.

THE DECISION TO INVEST IN A JOINT VENTURE OR A WHOLLY OWNED FACILITY

The company that chooses to invest in a facility to utilize its technology faces a further important decision: Should the investment be a joint venture with another institution, or should it be wholly owned by the technology provider? Some of the same factors that affected the licensing vs. investment decision were found to influence this choice as well. Specifically, the propensity of a firm owning technology to enter into a joint venture as compared with a wholly owned facility was greater:

- When the transfer was an international one (in fact, joint ventures were seldom used for domestic investments);
- When the technology was transferred to a developed, large-market country; and
- When a relatively small company was making the decision.[12]

The other variables—the number of companies owning the technology and the nationality of the firm—did not seem to affect the decision. Thus, in some respects—but not all—the degree of ownership seems to be an extension of the licensing vs. investment decision, as some of the factors that favor licensing over investment also favor joint ventures over wholly owned facilities.

RESULTING PATTERNS OVER TIME

The histories of the Nine Products suggest that patterns of technology transfer fall into three distinct phases. During the first period, which lasted for a decade or two in each case, the larger manufacturing firms dominated the business. They developed their own technology and used it in facilities wholly owned by them and located in their own home countries. The number of international transfers of technology was virtually nil, and the number of sales of technology was very low.

During the second period, which lasted for another decade or two, the smaller manufacturing firms and engineering contractors began to develop their own technology. The smaller manufacturing firms employed their technology primarily in their own facilities located in their home countries. The engineering contractors, of course, sold technology, both at home and abroad. The large manufacturing firms, in the meantime, began to invest abroad and to license at home and abroad. During this period, technology owners sold their technology about as often as they used it in their own facilities, with domestic transfers favoring investment and international ones favoring licensing. Of the transfers that involved investment by a technology-owning firm, some two-thirds of the international ones were joint ventures, whereas virtually none of the domestic ones were.

During the third period, which corresponded roughly to years thirty through fifty in the product life cycle (see note to figure 8–1), the number of transfers by large firms continued to increase; but they were surpassed by the number of transfers by the smaller manufacturing firms and the engineering contractors. These latter two categories of firms accounted for most technology transfers both at home and abroad. The engineering contractors, of course, concentrated on selling technology, whereas the smaller manufacturing firms both sold technology and used it in their own facilities. During this period, some two-thirds of the technology transfers were international ones; and licensing became the dominant channel, accounting for almost 90 percent of the international transfers and slightly over half of the domestic transfers. Of the transfers that involved investments by a technology owner, joint ventures continued to account for some two-thirds of the international transfers but for very few of the domestic transfers. (Figure 8–3 gives a summary for all three periods.)

Some observers have expressed a fear that a relatively few multinational enterprises will dominate many industries in the world.[13] Although it is conceivable that the petrochemical industry will eventually enter a fourth period, with entirely different ownership patterns, the histories of individual petrochemicals chronicle a strong decline in the economic power of the manufacturing firms that own technology— and these are principally multinational enterprises. For the Nine Products, for example, only 30 percent of international technology transfers involved investments by the technology owner during the second period, and only 14 percent during the third period. Most of these investments were joint ventures.

This loss of control by multinational enterprises also extends to manufacturing operations in the less developed countries. At the end of 1974, there were fifty-three plants in less developed countries manufacturing the Nine Products. Of these, only 15 percent were wholly owned subsidiaries of a foreign firm, and another 37 percent were joint ventures between local firms and foreign firms. Thus, the largest single block—48 percent—had no foreign ownership.

But technological dependence is a separate question. In all cases, the petrochemical technology used in plants in the less developed countries came from one of the developed countries. True, some less developed countries had reached the stage at which their internal sources could provide engineering and construction know-how for

Fig. 8–3. Transfers of technology via different channels, during three time periods in products' lives, domestic compared with international, nine representative petrochemicals, worldwide, through 1974.

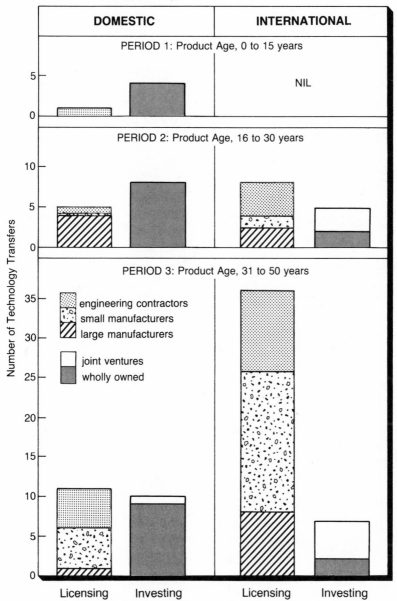

SOURCE: Industry trade journals plus author's questionnaire.

auxiliary facilities, and some were beginning to export this expertise in minuscule quantities. In a 1977 survey of then-current petrochemical projects, eight of the 580 engineering and construction firms involved in some phase of the project were headquartered in less developed countries; and one of the 223 exports of engineering and construction services was made by a firm headquartered in a less developed country.[14] But these beginnings of petrochemical-related know-how were still concentrated in the older engineering fields, especially civil and, to a lesser extent, mechanical. The development of process technology, which relies heavily on chemical engineering—one of the newer engineering fields—had barely begun in less developed countries.

Countries that import and export technology in the petrochemical industry can be arrayed along a sort of "technological ladder" (see figure 8–4). The bottom rung represents a stage in which a country

Fig. 8–4. Technology ladder of nations in the petrochemical industry.

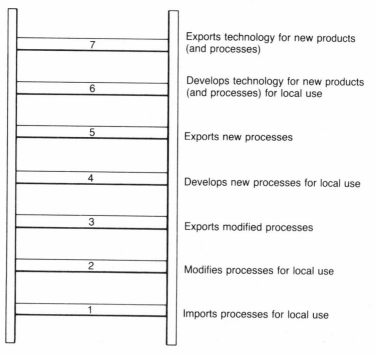

7 Exports technology for new products (and processes)

6 Develops technology for new products (and processes) for local use

5 Exports new processes

4 Develops new processes for local use

3 Exports modified processes

2 Modifies processes for local use

1 Imports processes for local use

imports the entire process. At the top—the seventh rung—the nation is an exporter of know-how for new products, including the processes used to make them. Of course, in actuality the rungs are not so clearly defined; there is much overlapping. The chart provides only a convenient way of thinking about national differences in regard to international trade in petrochemical know-how.[15]

In summary, the market for petrochemical technology changes dramatically as a product becomes older. In the early years, the technology is held tightly by a monopolist or a relatively few oligopolists. But after a number of years (a decade or two for the Nine Products), the tight oligopoly begins to break down and the licensing of technology becomes more widespread, particularly with the beginning of licensing by engineering contractors. After some more time passes (another decade or two for the Nine Products), imperfections in the market for technology are so eroded that licensing becomes the dominant channel for technology transfers. In this latter stage, competition in both technology sales and product manufacturing is at a sufficiently high level that the oligopoly profits derived from either of the two activities are likely to be relatively low. Indeed, this decline in oligopoly profits is reflected in the prices obtained for individual petrochemicals.[16]

APPENDIX

STATISTICAL TESTS OF MULTIPLE REGRESSION MODEL

Multiple regression techniques were used to model the transfer of technology by petrochemical manufacturing firms. Specifically, a firm's propensity to sell technology to an unrelated party rather than invest in its own manufacturing facility was hypothesized to be a function of a number of variables, as shown in the equation below:

$$\log_e \frac{p}{1 - p} = \beta_0 + \beta_1 X_1 + \beta_2 X_2 + \beta_3 X_3 + \beta_4 X_4 + \beta_5 X_5 + \beta_6 X_6 + \beta_7 X_7$$

where:

p = "true" but uncertain long-run fraction of technology transfer that took place via sale to unrelated party rather than to a facility wholly or partially owned by the company transferring the technology, among technology transfers with specified values of X_1, X_2, X_3, X_4, X_5, X_6, and X_7, as defined below

X_1 = number of companies that were engaged in petrochemical manufacturing activities and that owned a commercial process to manufacture the product in question

X_2 = number of companies that were not engaged in petrochemical manufacturing activities (all were engineering contractors) but that owned a commercial process to manufacture the product in question

X_3 = size of company transferring technology, as a percentage of size— measured by sales—of largest firm in the industry, where industry is either oil or chemical (data available only for U.S.-based petrochemical manufacturing firms)

X_4 = size of market of nation to which technology was transferred (manufacturing output of nation in billions of dollars in 1958)

X_5 = level of economic development of country to which technology was transferred (1 = developed, 0 = less developed)

X_6 = nation in which firm transferring the technology was headquartered (1 = U.S., 0 = not U.S.)

X_7 = domestic/international transfer (1 = technology transferred to facility located in home country of company transferring technology, 0 = technology transferred outside home country)

TABLE 8A-1

REGRESSION RESULTS

SAMPLE	$\bar{R}^2$	NUMBER OF OBSERVATIONS	CONSTANT (NAT. COEFF.)	U-STANDARDIZED (AT P = .5) COEFFICIENTS (Probability That Sign is Correct is Shown in Parenthesis)						
				X_1	X_2	X_3	X_4	X_5	X_6	X_7
All transfers	.29	447	-.42	.07 (.95)	.12 (1.0)	-.10 (1.0)	.17 (1.0)	.06 (.97)	-.01 (.60)	-.30 (1.0)
International transfers only	.21	213	.28	.01 (.56)	.17 (.99)	-.20 (1.0)	.15 (.94)	.06 (.90)	-.02 (.65)	

NOTE: The use of an interactive variable for X_1 and X_2 and the use of logs for variables X_1, X_2, and X_3 did not change the conclusions.

TABLE 8A–2

Correlation of Variables

Variable	Y Transfer Channel	X_1 No. of Manufacturers	X_2 No. of Engineering Companies	X_3 Source Company Size	X_4 Host Market Size	X_5 Host Development Level	X_6 U.S. Company	X_7 Home Country Location
				ALL TRANSFERS				
Y Transfer Channel	1.0							
X_1 No. of Manufacturers	.30[a]	1.0						
X_2 No. of Engineering Companies	.30[a]	.64[a]	1.0					
X_3 Source Company Size	−.10	−.11	.10	1.0				
X_4 Host Market Size	−.11	−.30[a]	−.18	.13	1.0			
X_5 Host Development Level	−.06	−.30[a]	−.17	−.09	.31[a]	1.0		
X_6 U.S. Company	.11	−.12	−.25[a]	.09	.48[a]	.15	1.0	
X_7 Home Country Location	−.42[a]	−.31[a]	−.27[a]	−.05	.55[a]	.32[a]	−.10	1.0

[a]Statistically significant at 95 percent confidence level.

9

Foreign Licensing in Multinational Enterprises

PIERO TELESIO

As the previous chapter pointed out, the principal channels used by multinational enterprises for the international transfer of technology are various forms of direct investment and licensing of unrelated parties.[1] While researchers have paid considerable attention to direct investments, little research has been devoted to licensing activities.[2] This study examines the licensing issues in depth by addressing four questions:

1. Why do multinational enterprises license their technology abroad?

This chapter is based on the author's "Foreign Licensing in Multinational Enterprises," D.B.A. dissertation, Harvard Business School, 1977, from which the book *Technology Licensing and Multinational Enterprises* (New York: Praeger, 1979) is derived.

2. What characteristics of a multinational enterprise are associated with a greater propensity to license?
3. How does competition affect licensing behavior?
4. Does the nationality of the multinational enterprise make a difference in licensing behavior?

Thus, it extends a portion of the study of petrochemicals reported in the previous chapter by examining the licensing issue for a wide range of industries.

To answer the four questions, a model of licensing was tested empirically. A total of sixty-six multinational enterprises participated in this study. Representatives of fifty-five of these firms were interviewed at length in person; the remaining eleven either responded to a mailed questionnaire or replied by letter. Of these sixty-six multinational enterprises, forty were based in the United States and twenty-six were based in Canada and Europe.[3] This study was limited to multinational enterprises primarily engaged in manufacturing; thus, any firms that had substantial extractive activities—for example, the large multinational oil companies—were excluded.

Foreign licensing is defined in this study as the sale of manufacturing technology by a multinational enterprise to a noncontrolled entity located outside the home country of the multinational enterprise. It is assumed that more than 50 percent ownership is needed to give the multinational enterprise control of a foreign operation; thus, this study includes only those operations abroad in which the multinational enterprise has a 50 percent or less ownership.

The study has several further limitations. First, it excludes licensing of proprietary rights not directly associated with manufacturing, such as licensing of trademarks or brand names. Second, it includes only those licensing agreements for which technology is purchased for money (direct payments or equity shares) or exchanged for other technology. This eliminates inactive licensing agreements, which might still be in effect but which generate no payments, and licensing by patent pools, in which members have free access to the patents held by the pool regardless of individual contributions. Third, the study is limited to technologies actually used by a multinational enterprise in its manufacturing operations at home or abroad; thus, the sale of technology not used by the seller is excluded. Fourth, also excluded are licensing of freely accessible technologies, such as bottling, packaging, or simple assembly operations; and agreements concerning management, engineering, or other nonmanufacturing services. Finally, the study also excludes licensing to communist countries.

WHY MULTINATIONAL ENTERPRISES LICENSE ABROAD

Licensing by multinational enterprises may be interpreted as an effort by managers to maximize benefits for the firm. The proprietary rights and knowledge relinquished in a licensing agreement are exchanged for certain assets of the licensee that the multinational enterprise values. Such a view implies that the decision to license technology to a noncontrolled entity abroad involves a rational evaluation of the benefits, costs, and risks of licensing as compared with the alternatives. This evaluation will depend in part on the strategies followed by the multinational enterprise and the resources it possesses; thus, multinational enterprises with different strategies and resources will evaluate the net benefits of a licensing agreement differently.

This study examines the impact on licensing decisions of a strategy of product or process innovation and of a strategy of product line diversification. The resources available to the multinational enterprise—for example, capital, management personnel, knowledge of foreign markets—are indicated by its relative size in its industry and by its experience in manufacturing in foreign markets. In addition, since this study includes multinational enterprises based both in the United States and abroad, the effect of the home country on licensing is also considered. Another relevant factor is likely to be the competition faced by the multinational enterprise in selling its technology in specific foreign markets. Thus, of the six factors considered in this study, five are characteristics of the multinational enterprise and the other is a characteristic of the environment. Licensing decisions are surely also affected by other factors, such as the size and rate of growth of the local market, the availability of qualified licensees, or the level of political risk, but these are beyond the scope of the present study.

There are two broad reasons why a multinational enterprise would license abroad: to substitute for controlled foreign direct investment when licensing proves to be more profitable, and to gain access to technology of other firms through reciprocal grants of licenses. In the first case, licensing to a noncontrolled entity provides an alternative to entering foreign production with a controlled investment; in the second, access to other firms' technologies provides an alternative to R&D.

All of the multinational enterprises interviewed indicated that they consider licensing to noncontrolled entities as a substitute for controlled foreign direct investment.[4] Viewed as a substitute for investment, licensing can be placed in the context of the product life cycle model of trade and investment.[5] When a firm decides to begin production in a foreign market, its manufacturing operation can take several alternative forms. The owner of the technology may choose to invest with a wholly owned subsidiary or a controlled joint venture (defined in this study as one in which the technology provider holds more than 50 percent of the voting stock); alternatively, the technology owner may license to a noncontrolled foreign company or individual. The product life cycle model indicates only the timing of manufacturing abroad; it is the company contemplating foreign production that determines the *form* it takes.

Licensing, then, is undertaken if its balance of benefits against costs and risks is more favorable than that of controlled investment. Through licensing, a multinational enterprise can obtain royalties from foreign markets that it could not otherwise profitably exploit directly through controlled subsidiaries. If the firm lacks certain resources, a licensee can provide capital, managers, and local marketing experience.[6] Also, a licensee can provide manufacturing capabilities in markets that are too small for economical production by a multinational enterprise,[7] or in markets where foreign ownership of investment is restricted.[8]

On the other hand, a multinational enterprise might want full control over its international operations in order to manage production, marketing, and finance worldwide, so as to maximize benefits to the whole multinational system.[9] Unless all foreign operations are controlled, the company may not be able to carry out some of its policies, such as a policy of production rationalization involving subsidiaries in different countries. A multinational enterprise, for example, might want to maintain tight control over all of its manufacturing operations because of worldwide standardization of products and interchangeability of parts.

Multinational enterprises also license for reciprocal access to technology; but only twenty-three of the sixty-six multinationals studied indicated that they have licensed for this reason. In this case, the motive for licensing is to obtain the technology of other innovative firms through reciprocal exchanges of licenses. Many technologies may be acquired through simple, nonreciprocal licensing agreements, but others are not easily available unless valuable technology is offered in exchange, either at the time of the agreement or in the future.

In industries where the rate of technological advance is rapid and where innovations often build upon one another, the exchange of licenses among competitors prevents any one firm in the industry from blocking the technological development of the industry as a whole by refusing to license certain key technologies. In addition, a liberal licensing policy among innovative firms provides savings on the cost of innovation, by avoiding duplication of some research while at the same time reducing the risk of excluding any one firm from access to vital new developments. Multinational enterprises also seek reciprocity in licensing when they need technologies from competitors to fill gaps in their product spectrum, or when they want to enter new businesses. The understanding is that the competitor would expect an equivalent service if needed in the future.

In general, reciprocal licensing serves to preserve stability in technology-based oligopolies.[10] Reciprocal licensing is, in fact, a risk-reduction strategy, by which the members of the oligopoly are assured that no innovation will give one firm a substantial advantage over the others.

Cross-licensing has been examined in several recent studies.[11] Though it is seldom precisely defined, cross-licensing may generally be considered a special case of reciprocal licensing, with the exchange of licenses occurring concurrently and with payments, if any, netted out. In this study, therefore, cross-licensing is subsumed under reciprocal licensing and not considered a separate category.

The multinational enterprises that reported they license for reciprocal access to technology also indicated that they license to supplement their foreign direct investments. Their aggregate licensing activity, therefore, reflects a mix of motivations, and indeed particular licensing agreements might serve both goals at once. These enterprises are almost exclusively concentrated in the chemical and pharmaceutical industries (SIC 28) and in the electrical machinery industry (SIC

36).[12] In the United States, these industries spent more on R&D as a percentage of industry sales than the average for all manufacturing industries.[13]

A linear regression model was used to determine the differences in characteristics between multinational enterprises that license for reciprocity and those that do not.[14] As shown in table 9–1, multinational enterprises that license for reciprocity tend to have relatively higher R&D expenditures and to be relatively diversified. (The column labeled "probability of sign" indicates the probability that the sign of the coefficient is the true one. Probabilities of 0.90 or greater are considered here to be statistically significant.) No significant connection was found between a policy of licensing for reciprocity and three other company characteristics: the relative size of the firm in its main industry; the percentage of total sales manufactured abroad by controlled subsidiaries, which is assumed to be indicative of foreign manufacturing experience; and whether or not the firm is based in the United States.

TABLE 9–1

LINEAR MULTIPLE REGRESSION MODEL OF MOTIVATION FOR LICENSING

EXPLAINED VARIABLE: 1 IF COMPANY LICENSES FOR RECIPROCAL ACCESS TO TECHNOLOGY; 0 OTHERWISE.

EXPLANATORY VARIABLES	SIGN OF COEFFICIENT[a]	PROBABILITY OF SIGN[b]
R&D relative to sales	positive	1.00
Product diversification	positive	1.00
Relative size	positive	0.80
Percent manufactured abroad	negative	0.57
U.S. nationality	negative	0.67
Coefficient of determination	0.52	
Number of observations (of which 23 licensed for reciprocal access to technology)	65[c]	

SOURCE: Piero Telesio, *Technology Licensing and Multinational Enterprises* (New York: Praeger, 1979), Table 4–1, p. 53.
NOTE: See Appendix to this chapter for details of model.
[a]Sign of the beta coefficient of appropriate variable in regression model.
[b]Probability, in Bayesian terms, that the sign of the coefficient is as stated for the universe for which this sample could be considered random.
[c]One observation excluded because data on R&D expenditures were not available.

It is not surprising that multinational enterprises that license for reciprocity have relatively large R&D expenditures. Reciprocal licensing is thought to take place among firms that participate in technology-based oligopolies, where barriers to entry are erected by a rapid rate of technological innovation. To maintain this high rate of innovation, large expenditures on R&D as a percentage of sales are necessary.[15]

The multinational enterprises were divided into two categories: those that license *only* to supplement their foreign direct investments, and those that *also* license for reciprocal access to technology. The two types of firms have different strategies and characteristics; that is, those in the second category are concentrated in certain industries and have relatively large R&D expenditures and a high degree of product diversification. Moreover, since they use licensing in a double role, they have a complex licensing policy. Each category is considered separately.

LICENSING TO SUPPLEMENT FOREIGN DIRECT INVESTMENT

We began with a hypothesis that a firm's licensing activity is a function of its strategy and resources. Table 9–2 presents the results of a statistical analysis made in an effort to explain the behavior of multinational enterprises that license only to supplement foreign direct investment in terms of five variables indicative of strategies and resources:

1. R&D expenditure as a percentage of total sales, which is assumed to be a proxy for a strategy of innovation;
2. The degree of product diversification, a proxy for a strategy of diversification;
3. The relative size in industry, a proxy for resources available to the firm;
4. The percentage of total sales manufactured by controlled foreign subsidiaries, a proxy for foreign manufacturing experience; and
5. Whether the multinational enterprise is based in the United States.

TABLE 9–2

LINEAR MULTIPLE REGRESSION MODEL EXPLAINING CHARACTERISTICS OF MULTINA-
TIONAL ENTERPRISES THAT DO NOT LICENSE FOR RECIPROCITY

EXPLAINED VARIABLE: FOREIGN ROYALTY RECEIPTS AS A PROPORTION OF FOREIGN
SALES.

EXPLANATORY VARIABLES	SIGN OF COEFFICIENT[a]	PROBABILITY OF SIGN[b]
R&D relative to sales	positive	0.94
Product diversification	positive	1.00
Relative size	negative	1.00
Percent manufactured abroad	negative	0.96
U.S. nationality	negative	0.87
Coefficient of determination	0.58	
Number of observations	33	

SOURCE: Piero Telesio, *Technology Licensing and Multinational Enterprises*, Table 5–1, p. 68.
NOTE: See Appendix to this chapter for details of the model.
[a]Sign of the beta coefficient of appropriate variable in regression model.
[b]Probability, in Bayesian terms, that the sign of the coefficient is as stated for the universe for which
this sample could be considered random.

The amount of licensing is measured by royalty and fee receipts
from foreign licensees as a percentage of sales by controlled foreign
subsidiaries. This measure indicates the relative importance of licensing revenue as compared with revenue from controlled foreign direct
investment.[16]

Table 9–2 shows only the sign of the regression coefficients in the
statistical analysis because only the direction of the influence, not the
magnitude, is considered here. Again, probabilities above 0.90 are
considered statistically significant.

Expenditure on R&D. As shown in table 9–2, multinational enterprises that spend more on R&D as a percentage of sales license
relatively more than do firms that spend less on R&D. The strong
relationship is surprising. The development of technology that can be
licensed is often the result of R&D expenditures by the enterprise,[17]
but managers of multinational enterprises generally say that they
prefer to exploit important innovations directly through controlled
foreign operations.[18] However, certain technologies may be of only

marginal interest to the company and thus are licensed more readily. Moreover, in innovative firms, new technologies rapidly replace the old ones. These older technologies may then be licensed, since they have become less important to the firm's competitive strength. Furthermore, foreign firms seeking licenses will be likely to approach the more innovative companies. This creates more opportunities for licensing in innovative companies, resulting in a relatively greater licensing activity.

Product diversification.   Table 9–2 confirms that more diversified multinational enterprises license more than less diversified firms. In a diversified firm, managerial, production, and marketing resources must be allocated over many different product lines. This allocation is necessarily uneven because of the lumpiness of resources; and in any new operation—at home or abroad—there is a minimal investment in resources that must be provided for a viable business. The highly diversified multinational enterprise, therefore, cannot make the required investments for all of its product lines in all foreign markets where it wishes to operate, so it seeks additional resources from foreign licensees.

When a diversified firm wants to increase its product line diversification abroad, a licensee can offer rapid access to foreign markets. For example, each product line might require a distinct marketing formula, different for each country, which the firm could execute internally only at considerable cost. A licensee can offer its own marketing expertise, having already sunk costs into acquiring knowledge of its own market, training personnel, and developing sales channels.

In addition, the loss of control over technology, production, and marketing that is implicit in licensing is probably of less concern to diversified companies, because each product line accounts for only a small share of earnings and sales. Consequently, the highly diversified firm will not need to exercise full control over the performance of each product line in every market where the company is operating.

Furthermore, each product division competes with other product divisions for company resources. As a result, top management often imposes high return-on-investment standards.[19] Some divisions, therefore, may choose to license if an investment in a controlled foreign subsidiary would not meet the high standards set for the division.

As one executive interviewed summed it up: "If the size of the market is small, then license. We cannot devote management time to explore and manufacture in lesser markets."

Relative size.   Each multinational enterprise was categorized as small, medium, or large according to its relative size in its main industry. The regression in table 9–2 shows that relative size is a significant factor in determining a multinational enterprise's propensity to license: the larger the firm, the less licensing it undertakes.

Past studies have shown that a follow-the-leader pattern is common among multinational enterprises investing abroad.[20] One company, often one of the larger firms in its industry, leads with an investment in a foreign market. Within a short time, other firms in the same industry follow with their own investments. In addition, supplier firms, large or small, usually have to follow their customers in their foreign investments if they want to retain business abroad.

The smaller multinational enterprises in an industry, therefore, often have to match the foreign investments of the larger ones. The smaller firms, however, do not have the resources and cannot expand fast enough to keep up with the foreign investments of large competitors or of the firms they supply. Since resources often come in indivisible packages, the smaller firm that wants to be present in a number of foreign markets cannot allocate its capital and management resources uniformly over all of its foreign operations and must depend on the additional resources offered by local licensees. As a result, the first step for a relatively small multinational enterprise entering a foreign market is often to license a noncontrolled local firm. This move requires very little in the way of resources from the multinational enterprise other than the technology itself.

One small multinational enterprise in transportation equipment—a supplier to other firms in its industry—mentioned five reasons for licensing abroad: 1) licensing does not require capital investment; 2) licensing does not tie up qualified personnel on a long-term basis; 3) licensing generates revenue immediately; 4) licensing creates the basis for an investment opportunity; and 5) licensing provides an opportunity to get to know the people you are dealing with.

Experience in manufacturing abroad.   Table 9–2 also indicates that, the greater the share of sales of products manufactured abroad, the smaller the relative importance of licensing. Manufacturing experience in foreign markets is a precious resource, acquired only at the price of doing it. The more operations a multinational enterprise has abroad, the more the experience of operating in different environments becomes internalized as a "technology" of multinational operations. Multinational enterprises that have little of this operating know-how will be likely to value highly the managerial, marketing, and production capabilities of local licensees. The uncertainties of entering new foreign markets with a manufacturing operation might seem quite forbidding to such a firm. Other studies also have shown that firms with relatively little foreign experience are more likely to license than to invest.[21] One company that manufactured less than 10 percent of total sales through controlled subsidiaries abroad stated in an interview that it "always considers licensing as the first step to entering a foreign market with local manufacturing."

National base of the multinational enterprise.   The multinational enterprises were identified as being either based in the United States or not; a finer classification by nationality was not warranted by the sample size. As shown in table 9–2, regression did not indicate a statistically significant relationship between U.S. nationality and a multinational enterprise's propensity to license. This result is consistent with earlier studies concluding that, although there are some important and visible differences, multinationals based in the United States and multinationals based in Europe exhibit similar responses to common problems.[22] Indeed, this similarity has been found to extend explicitly to decisions concerning the ownership of foreign subsidiaries. One study concludes that, when other firm characteristics are similar, "it is difficult to substantiate any great difference in ownership policies between American and European companies."[23] Thus the finding that nationality seems not to make an important difference in licensing policy is not surprising, since some motivations for licensing are akin to those for taking on joint-venture partners.

Competition.   Managers have been advised to introduce new products and processes abroad through wholly owned subsidiaries. But as these products and processes mature and competition increases, managers have been advised to enter into joint ventures or to license.[24]

Competition can affect the decision to license in two ways. First, a relatively large number of competitors in a specific market generally means lower prices, revenues, and profits for individual firms. This makes investment with a controlled subsidiary relatively less attractive and licensing relatively more attractive[25] (provided, of course, that the reduced control over foreign operations implicit in licensing does not lower the profitability of the entire company). An investment may even have a negative return, while licensing generates a positive cash flow with a minimum of investment.

Second, if several firms are willing to sell competitive technologies, licensing again becomes relatively more attractive as a means of market entry. For example, if a foreign market is large enough to support a local manufacturing operation, a firm can choose between licensing and investing with a controlled subsidiary. If competitive technologies are available from several sources, a prospective local licensee might license from another source should the firm itself choose to invest. The licensee might then be able to preempt the market (since such a firm would generally need to make only a marginal, as opposed to a full, investment in order to begin production). When there is competition in the sale of technology, firms typically take the attitude that "if we do not license, somebody else will."

True, competition in the sale of technology also lowers the price of the technology, resulting in lower royalty payments. But in such an environment, a full investment might result in a return below that considered acceptable, or even in a negative return. In such a case, licensing would be preferable, even if royalty returns are low.

The effect of competition on licensing was not tested by means of a multiple regression, because data for this study were obtained for the firm as a whole rather than for individual products; and a company-wide measure of competition faced by the firm is not appropriate. Suppose, for example, that a company has one product that faces keen competition and that it licenses technology for this product but uses wholly owned facilities to produce a large number of products for which competition is less intense. An aggregate measure of the competition faced by the firm would obscure the fact that the firm *does* license if a particular technology has competitors.

Managers of the multinational enterprises in the sample were asked if they would be more willing to license a product or process if the number of competitors were large; fourteen replied yes, and eleven replied no. (Only twenty-five firms answered this question.) The aggregate result is not very illuminating, although it does indicate that the presence of competition increased the willingness of some managers to license.

It is more helpful to examine separately the responses from firms that have low, medium, or high product diversification. Table 9–3 shows that, when competition is present, the willingness of managers to license is greater for the more diversified firms. All of the managers of firms in the high-diversification category answered that they would license more readily if there were competition, as compared with 50 percent in the medium category and only 33 percent in the low category.

Highly diversified firms generally do not erect marketing and production barriers to entry in order to preserve their maturing products from competition, as might be the case for a firm with only one basic product line.[26] Thus these highly diversified firms are likely to have a number of older, mature products facing price competition, and these products are candidates for licensing.

TABLE 9–3

QUESTIONNAIRE RESPONSES ON COMPETITION, FROM MULTINATIONAL ENTERPRISES THAT DO NOT LICENSE FOR RECIPROCITY, CLASSIFIED BY DEGREE OF PRODUCT DIVERSIFICATION

| | PRODUCT DIVERSIFICATION | | | | | |
| | Number | | | Percent | | |
	Low	Medium	High	Low	Medium	High
License more readily if competition?						
Yes	3	5	6	33	50	100
No	6	5	0	67	50	0
Total number of companies	9	10	6	100	100	100

SOURCE: Piero Telesio, *Technology Licensing and Multinational Enterprises*, Table 5–12, p. 88.
NOTE: There is a .99 probability that the absolute difference between the "medium" and "high" categories is greater than 0 in the universe for which this sample could be considered random. The probability that there is a greater than 0 difference between the "low" and "high" categories is 1.0. See Appendix to this chapter for details.

Licensing for Reciprocity

Multinational enterprises that license for reciprocity also license to supplement their foreign direct investments, so that their total licensing activity is a combination of both types. One would expect their licensing to supplement foreign direct investment to be affected by the same characteristics that influence licensing in firms that do not license for reciprocity. For example, a relatively small multinational enterprise in the chemical industry (an industry in which licensing for reciprocity is practiced) should be more likely to license abroad to supplement its foreign direct investments than would a large firm in this industry.

Table 9–4 shows the linear regression model for licensing for multinational enterprises that have both types of licensing. The explained variable is the same as in the previous regression model except that the foreign royalty receipts from licensees include revenues from both types of licensing; these receipts, as previously, are taken as a proportion of foreign sales. The explanatory variables, indicating certain characteristics of the multinational enterprise, are also the same as for the other regression model.

TABLE 9–4

LINEAR MULTIPLE REGRESSION MODEL EXPLAINING CHARACTERISTICS OF MULTINATIONAL ENTERPRISES THAT LICENSE FOR RECIPROCITY

EXPLAINED VARIABLE: FOREIGN ROYALTY RECEIPTS AS A PROPORTION OF FOREIGN SALES

EXPLANATORY VARIABLES	SIGN OF COEFFICIENT[a]	PROBABILITY OF SIGN[b]
R&D relative to sales	positive	0.84
Product diversification	positive	0.93
Relative size	positive	0.65
Percent manufactured abroad	negative	1.00
U.S. nationality	negative	0.97
Coefficient of determination	0.63	
Number of observations	20	

SOURCE: Piero Telesio, *Technology Licensing and Multinational Enterprises,* Table 6–1, p. 96.
NOTE: See Appendix to this chapter for details of model.
[a]Sign of the beta coefficient of appropriate variable in regression model.
[b]Probability, in Bayesian terms, that the sign of the coefficient is as stated for the universe for which this sample could be considered random.

The regression results for firms that license for reciprocity are not identical to those for the other firms. It is not clear whether the differences are due to a small sample size (only twenty observations for firms that license for reciprocity) or whether they actually reflect systematic differences in the behavior of the two types of firms.

Similar results were obtained for two of the explanatory variables. For both types of firms, the propensity to license increases with the firm's diversification and its experience in manufacturing abroad. But the regression analysis gave dissimilar results for the other three explanatory variables.

First, consider research and development expenditures. As suggested by the earlier analysis presented in table 9–1, licensing for reciprocity was hypothesized to be a linear function of the firm's expenditure on R&D relative to sales; that is, the greater this expenditure, the more licensing for reciprocity the firm undertakes. Similarly, as indicated in table 9–2, increasing R&D expenditures also increase the propensity to licensing as a supplement to foreign direct investment. Since the firms examined here participate in two kinds of licensing, both positively associated with increasing R&D expenditures, one would expect their licensing activity to increase with increasing R&D. But the regression analysis did not confirm this expectation (see table 9–4). The coefficient of the R&D variable, though positive as expected, is not statistically significant (having a .84 probability as compared with the .90 assumed to be significant). Those companies that license for reciprocity, of course, spend on the average significantly more on R&D as a percentage of sales than do companies not licensing for reciprocity. The regression analysis suggests that, within this group, relatively greater expenditures on R&D do not seem to result in a greater amount of licensing.

Moreover, contrary to findings for companies that do not license for reciprocity, relative size does not appear to affect the propensity to license of firms that *do* license for reciprocity. The sign of the regression coefficient in table 9–4 is positive but is not statistically significant. This finding might be the result of two offsetting trends. It is possible that the amount of licensing for reciprocity *increases* with increasing relative size, even as licensing to supplement foreign direct investment *decreases*. In such a case, the sum of the two types of licensing would not show any clear effect of size. Licensing for reciprocity may increase with relative size because it is the highly innovative firms that are most likely to license for reciprocity; some past studies have found a positive correlation between innovation and size.[27]

The nationality of the parent is the other variable for which the results of this analysis are not completely consistent with earlier results for firms that do not license for reciprocity. Both analyses suggest that U.S.-based multinational enterprises would be less likely to license than would non-U.S. firms, but the coefficient for this variable was statistically significant (probability of .97) for the analysis of firms that practice reciprocal licensing and not statistically significant (probability of .87) for firms that do not license for reciprocity. If indeed this is a true difference between these two samples, it might be explained by a greater propensity on the part of non-U.S. firms to license for reciprocity. U.S. antitrust laws could have some deterrent effect on U.S. firms in this respect.

Competition also affects licensing policy in multinational enterprises that license for reciprocity. In response to the question, "Would you more readily license a product or process if it faced competition?" fourteen managers of these companies answered yes, and five answered no. These responses cannot be directly compared with those of firms licensing only to supplement foreign direct investment, because the degree of diversification affected the decisions of this group of firms. Of the firms that license for reciprocity, sixteen have high product diversification and the remaining three have medium diversification. Therefore, to obtain a comparison of companies with an equal degree of diversification, only the responses of the companies with high product diversification are used. All six of the highly diversified companies that do not license for reciprocity (and that responded to this question) stated that competition increased their willingness to license. On the other hand, only twelve out of sixteen highly diversified companies that license for reciprocity gave this reply. (The difference in distributions is statistically significant, with a probability of 1.00.) This indicates that, for the latter type of firms, competition perhaps has somewhat less of an effect on overall licensing decisions.

The responses of managers of the companies that license for reciprocity reflect mixed motivations for licensing. While licensing to supplement foreign direct investment is generally affected by the presence of competition, especially for highly diversified firms, licensing for reciprocal access to technology is probably not, because reciprocal licensing decisions do not usually involve profitability comparisons with a proposed direct investment. In addition, some technologies are licensed for reciprocity before any competition has developed. For example, some innovations in semiconductors are licensed almost immediately, in order to prevent stalemate situations in the industry.[28]

SUMMARY OF FINDINGS

This study finds that multinational enterprises, whether based in the United States or not, view licensing to noncontrolled entities abroad as an alternative to controlled foreign direct investment. The motivations for licensing are to establish a manufacturing presence in a country and to earn a more attractive return for the multinational enterprise system than would have been possible with a controlled manufacturing facility.

Licensing also is sometimes undertaken to gain access to the technology of other innovators through reciprocal exchanges of licenses. Multinational enterprises engaging in this practice, in comparison with other multinational enterprises, tend to spend more on R&D as a percentage of sales and are more diversified. They are concentrated in the pharmaceutical, chemical, and electrical industries. Thus there appear to be two categories of multinational enterprises: those that license only to supplement their foreign direct investments, and those that engage in reciprocal arrangements to obtain technology. These two categories of firms were examined separately.

For multinational enterprises that license only to supplement their foreign direct investments, the amount of licensing was found to be greater for companies that spend more of sales on R&D, are more diversified, are among the smaller firms in their industry, and have relatively less experience with foreign operations. Conversely, companies that have a low propensity to license spend less on R&D as a percentage of sales, are less diversified, are among the large firms in their industry, and have a relatively greater experience in operating abroad. On the other hand, a firm's propensity to license does not seem to depend on whether it is based in the United States or not.

The licensing behavior of multinational enterprises that also license for reciprocal access to technology appears to be more complex. Although it was expected that firms spending more on R&D as a percentage of sales would license more (for both motivations), this expectation was not confirmed. Moreover, the larger firms in their industry are not likely to license less, as expected. This may be because the more innovative firms, which are likely to be the larger ones, do more reciprocal licensing. As expected, highly diversified firms and firms with less foreign operating experience are likely to license relatively more. An unexpected finding was that multinational enterprises based outside of the United States appear to license relatively more, perhaps because of national differences in practices regarding reciprocal licensing agreements.

On the average, competition in a product's market and in the supply of technology to manufacture the product appears to increase the willingness of managers of multinational enterprises to license in order to supplement their foreign direct investments. This behavior seems to apply especially to highly diversified companies, which generally are less likely than single-product-line companies to try to preserve their market position in older products as competition increases. On the other hand, the presence of competition appears to have a lesser effect on the willingness of managers to license for reciprocity.

CONCLUSIONS

Implications for Licensors

This study should be of interest to managers of multinational enterprises seeking to define a licensing policy that does not conflict with other company policies. Coordination with these other policies is necessary if the full benefits of licensing are to be realized. Licensing should not be pursued independently of foreign investment policy or, for innovative firms, independently of R&D policy. Licensing can play a very useful role by supplementing these policies rather than conflicting with them. Indeed, a policy of avoiding licensing at all costs can be damaging to the competitive position of a firm that would benefit from technological exchanges through reciprocal licensing.

In particular, managers of smaller multinational enterprises might wish to consider the advantages of entering foreign markets through licensing. In this way, the small multinational enterprise could spread its resources over more foreign markets than it could otherwise enter. As it grows, however, it might want to establish controlled subsidiaries in those markets. The small multinational enterprise need not avoid licensing to noncontrolled entities, since clauses can be written into licensing agreements to give the licensor an option to buy part or all of the licensee, thus assuring permanence to the licensing arrangements. A firm planning to establish a long-term presence in foreign markets through licensing might also consider licensing its trademarks. In this way, the market is developed for the licensor's, not the licensee's, trademarks.

Multinational enterprises with less experience operating abroad might also find licensing initially desirable. As these companies gain experience in foreign markets, however, they might want controlled operations, especially if they have low product-line diversification. Also, as they grow abroad, firms should plan to reduce their foreign licensing operations if they cannot be conveniently integrated with other foreign activities. Again, these companies should be careful to build up the licensee's market with their own, rather than the licensee's, trademarks.

Managers of multinational enterprises pursuing a strategy of product diversification may find licensing a useful way to introduce their products quickly into foreign markets. Specifically, firms that are both innovative and highly diversified can devote their resources to introducing new products rather than to investments in disparate markets. Licensing also provides an additional source of revenue for older products that the diversified multinational enterprise might otherwise discard from its product line. If such a company continuously introduces new products into its foreign markets, licensing agreements need not guarantee a permanent presence. Such a firm would probably have little interest in acquiring equity participation in licensees to ensure a long-term partnership. In addition, managers of diversified firms are less concerned than are managers of nondiversified firms with operating their foreign manufacturing activities as a wholly integrated system, as evidenced by the fact that these firms have relatively fewer wholly owned subsidiaries.[29] Thus, licensees can be easily tolerated, although this conclusion might apply more to a minor product line than to a major product line of a diversified company.

Implications for Licensees

Prospective licensees will probably have little success in obtaining technology from the larger firms in an industry or from firms with a low degree of product diversification. Companies seeking to purchase technology should concentrate their efforts on approaching smaller firms, more diversified firms, and firms with relatively little foreign manufacturing experience. They should, however, be aware that licensing from a small firm or from a firm with little foreign manufacturing experience may not be a stable arrangement over the long run, especially if the licensor has a narrow product line. Such licensors may eventually want to acquire the licensee or to set up a competing subsidiary while simultaneously terminating agreements with the licensee. Licensing agreements are probably more stable with highly diversified firms. If these firms grow, their policies toward licensing abroad tend to change less than those of firms with few product lines.

Licensees may find that the ability to offer technology in return is necessary to obtain access to state-of-the-art technology in the pharmaceutical, chemical, and electrical industries. A licensee wanting to purchase such technology should be prepared to do research at the same advanced level and to share some of the fruits of this research, through licensing, with competitors.

Implications for Governments

If governments restrict the outflow of foreign direct investment, the outflow of licensing will increase unless also restricted. Licensing can partially substitute for foreign direct investment in the transfer of technology abroad. However, the technology owner is likely to realize lower returns from licensing undertaken because of restrictions on foreign direct investment than from a controlled foreign direct investment.

If, on the other hand, licensing is restricted and foreign direct investment is not, it is not likely that foreign direct investment will increase significantly, because it typically is not a substitute for licensing. For example, if a company chooses to license because an investment would not be profitable, it would be unlikely to undertake the unprofitable investment if licensing were restricted.

Licensing restrictions would hurt firms by reducing royalty income. In addition, technology from competitors would most likely replace restricted U.S. technology in third-country markets. For example, a relatively small U.S. multinational enterprise, if not permitted to license, might have to forego establishing a manufacturing presence in foreign markets; direct investment would probably be too costly. Similarly, a U.S.-based multinational enterprise with high product diversification that licenses older technology would most likely abandon some markets to non-U.S. competitors and thus lose royalty income. In both of these examples, the chances are that technology would still be transferred, but from a source other than the United States, with the U.S.-based multinational enterprise losing both market presence and income and the United States losing taxes on royalties. In addition, licensing restrictions would produce indirect effects for both the firm and the nation involving a long-term loss of competitive strength vis-à-vis foreign competitors.

This analysis is consistent with the finding that U.S. firms generally invest abroad for defensive reasons, as they would otherwise lose the foreign markets to non-U.S. competitors.[30] And restrictions on U.S. foreign direct investment would be damaging to the United States in terms of the balance of payments and the skill levels of employment.[31]

Restrictions on reciprocal licensing would succeed in keeping certain innovations at home but could prevent firms from acquiring important foreign technology. Innovative firms might then fall behind the technology of firms in other countries that do allow licensing. If the United States were, for example, to impose such restrictions, U.S. firms in the pharmaceutical, chemical, and electrical industries would probably have access to fewer new technologies, as they would be deprived of the benefits of R&D carried out by other firms abroad. With a deteriorating competitive position, they might then return fewer benefits to the United States. Of course, foreign firms would fall behind the U.S. firms in certain fields, so they too would return fewer benefits to their home countries.

Countries that want to restrain the influx of foreign investment will often find licensing an attractive alternative. Still, government policy makers should be realistic with regard to the technology they can obtain through licensing. Often, certain technologies are not available for licensing without control, especially if the licensor is a large single-product-line firm. For example, it would not be easy to convince an automobile producer seeking area-wide production integration to operate in one country through a licensee. Moreover, developing countries are likely to obtain only limited access to advanced technologies in the pharmaceutical, chemical, and electrical industries, since it is unlikely that they could offer valuable technology in exchange.

Finally, while Japan provides a striking example of how a country can make use of foreign licensed technology,[32] not every nation will be able to match its performance. Developing countries, for example, sometimes do not have the capabilities to use manufacturing technology without the transfer of management skills associated with foreign investment.[33] For these countries, a policy of denying foreign ownership of investments might result in denying the country access to needed technologies.

APPENDIX

Table 9–1 in the text shows the regression results on the entire sample with the explained, or dependent, variable as a dummy variable with the value of 1 if a multinational enterprise licenses for reciprocal access to technology and a value of 0 if it does not. (Only sixty-five observations were included because R&D expenditures were not available for one firm.)

The explanatory, or independent, variables indicative of company characteristics are:

1. *R&D relative to sales:* R&D expenditures as a percentage of total consolidated sales.
2. *Product diversification:* degree of product-line diversification, where low = 1, medium = 2, and high = 3. A firm is classified as low if the sales of its principal product line (defined according to three-digit SIC) account for more than 90 percent of total sales, no matter how many other product lines it manufactures; or if the main product line accounts for more than 80 percent of total sales, but less than 90 percent, and it manufactures no more than two major product lines. A firm is classified as high if the principal product line accounts for no more than 60 percent of total sales and the firm manufactures in four or more product lines. All firms falling in between are classified as medium.
3. *Relative size:* relative size of the firm in its main industry, where small = 1, medium = 2, and large = 3. A firm is classified as large if total sales are greater than one-half of the total sales of the largest firm in its main industry; medium if total sales are between one-half and one-quarter the total sales of the largest firm; small if total sales are less than one-quarter of the total sales of the largest firm.
4. *Percent manufactured abroad:* percentage of total consolidated sales manufactured by controlled foreign subsidiaries.
5. *U.S. nationality:* national base of the multinational enterprise, where 1 = U.S.-based, 0 = other.

The regression model in table 9–1 is of the form:

Firm licenses for reciprocal access to technology = $C_0 + C_1$ (R&D relative to sales) + C_2 (product diversification) + C_3 (relative size) + C_4 (percent manufactured abroad) + C_5 (U.S. nationality).

The total sample of sixty-six multinational enterprises then was divided into two subsamples: one of forty-three multinational enterprises that license *only* to supplement foreign direct investment, and one of twenty-three firms that *also* license for reciprocity. These two subsamples were used to test models of licensing that relate linearly the relative importance of licensing to certain characteristics of the multinational enterprise.

The relative importance of licensing was measured by the royalty and fee revenue from foreign licensees as a proportion of total revenue (i.e., sales) from controlled foreign manufacturing subsidiaries. This is the explained variable.

Table 9–2 in the text presents regression results for thirty-three of the forty-three multinational enterprises that license *only* to supplement foreign direct investment (royalty receipts were not available for the other ten firms). The regression model is of the form:

> Relative importance of licensing = B_0 + B_1 (R&D relative to sales) + B_2 (product diversification) + B_3 (relative size) + B_4 (percent manufactured abroad) + B_5 (U.S. nationality).

For reasons given in the text, a measure of competition was not included in the multiple regression analysis; instead, a separate analysis was made for this variable. Table 9–3 in the text shows the number and relative frequency of responses, for the sample of multinational enterprises that license *only* to supplement foreign direct investment, to the question whether managers would license more readily if there were competition in the supply of technology. For this analysis, the firms were classified by the degree of product diversification. A total of twenty-five multinational enterprises out of forty-three replied to this question.

Table 9–4 in the text presents regression results for the multinational enterprises that license for reciprocity as well as to supplement foreign direct investment. Royalty receipts were available for twenty of the twenty-three firms in this category. The regression model is based on the hypothesis that the relative importance of licensing done by these firms is the sum of licensing done for reciprocal access to technology and licensing done to supplement foreign direct investment. The relative importance of licensing for reciprocal access to technology, in turn, is hypothesized to be a linear function of R&D expenditures as a percentage of sales:

> Relative importance of licensing for reciprocity = A_0 + A_1 (R&D relative to sales).

This is added to the model for licensing to supplement foreign direct investment to obtain:

Relative importance of licensing for reciprocity + relative importance of licensing to supplement foreign direct investment = $D_0 + D_1$ (R&D relative to sales) + B_2 (product diversification) + B_3 (relative size) + B_4 (percent manufactured abroad) + B_5 (U.S. nationality).

Therefore, we have:

$D_0 = B_0 + A_0$ and $D_1 = B_1 + A_1$.

10

Joint Ventures and the Transfer of Technology: The Case of Iran

FARSHAD RAFII

Whatever the attitudes of multinational firms toward wholly owned subsidiaries, joint ventures, and licensing arrangements, most governments have their own interests. But views as to what they should prefer are not unanimous.

On one side of the controversy, some writers argue that multinational enterprises, while pursuing their own objectives, not only transfer modern production technologies and managerial skills but also supply much-needed capital and material inputs, develop and utilize local resources in an efficient manner, provide access to export markets, and in general act as "engines of development"

This chapter is based on the author's "Joint Ventures and Transfer of Technology to Iran: The Impact of Foreign Control," D.B.A. dissertation, Harvard Business School, 1978.

wherever they go.[1] The proponents of this view advocate an open-door policy on the part of the developing countries and consider restrictions on foreign investment inimical to the host's own interests. The foreign investor, the argument would go, should be free to exercise any degree of control over investments.

An equally outspoken group of thinkers, on the other hand, maintains that the objectives pursued by the multinationals often conflict with the economic and political interests of host nations. Foreign investors may transfer little technology and know-how, stifle local entrepreneurial initiative, and create harmful dependencies and many other intended or unintended social and economic ills. Furthermore, it is argued, multinationals exact unreasonably high prices for whatever benefits they do provide by maintaining a high degree of control over their affiliates through various means.[2] According to this line of reasoning, the foreign supplier of technology cannot be left to its own devices.

Aware of arguments on both sides of the controversy, many developing countries have adopted a policy of requiring foreign investors to form joint ventures with local private or government entities. Since the interests of local partners are to be more in line with the overall interests of the host country than those of foreign firms, participation of local investors in the joint venture's decision-making process is expected to enhance the net benefits to the host country. Moreover, active involvement of nationals in day-to-day operations of a joint venture is believed to lead to a speedier transfer of know-how than would be the case with a wholly owned subsidiary.

Whether these policies are effective, however, is still an open question. It is possible that restrictions on ownership encourage the parent firm to withhold from the joint venture resources that it would provide freely to a wholly owned subsidiary. Moreover, ownership is not necessarily synonymous with control. The foreign firm may be able to exercise considerable influence over joint-venture operations without full, or even majority, ownership. Control of key managerial functions or contractual agreements that obligate the joint venture to behave in certain ways may be sufficient substitutes for ownership.

The objective of this study was to explore the relationship between the degree of foreign influence (ownership, managerial control, and contractual control) over a joint venture, and the extent and cost of technology and know-how transferred to the host country via the joint venture.

A joint venture was defined broadly to include pure licensing operations in which foreign equity is reduced to zero. The extent' of domestic value added, the degree of manpower indigenization, and the efficiency of the venture's operations were used as measures of the extent of technology transfer. The cost of the technology to the host was measured primarily through royalty payments and any excess prices paid by the subsidiary to the parent for imported intermediate products.

The field research was conducted in Iran in 1977. Data were obtained from thirty-five joint ventures in the manufacturing sector (nine automotive; eleven rubber, glass, synthetic fibers, petrochemical, and pharmaceutical; and fifteen metal forming, electrical, and nonelectrical). Three firms were foreign majority owned, three were fifty-fifty joint ventures, four were pure licensing operations, and the remainder were joint ventures in which the foreign partners held a minority share.

FOREIGN CONTROL AND DOMESTIC CONTENT

Whatever the ownership structure of a venture, local production is likely to start with the assembly of components imported mainly from sources affiliated with the foreign partner. The host government, however, generally seeks to induce the joint venture to expand its level of operations to include locally produced raw materials and components.

The local and foreign partners are both likely to resist pressures to increase domestic content, for a variety of reasons. With a characteristic "trader's mentality," each is likely to view the venture merely as a way to circumvent the trade barriers rather than as a genuine move to undertake local production.[3] If, as is normally the case in most developing countries, local sources of components and raw materials are scarce, then increasing local content means expansion of the joint venture's own operations, requiring additional capital investments. The foreign partner may be reluctant to sink additional funds into an uncertain environment. The local partner may also shy away from increasing his investment, either because he simply is unable to raise the required funds or because he prefers to avoid a large, fixed investment in a single operation.[4]

Even if the required inputs are available locally, the private partners are still likely to prefer importing, because imports are normally perceived as having higher quality, more reliable delivery, and lower cost than locally supplied goods. Furthermore, a large import volume permits convenient transfer of funds abroad by appropriately pricing imported goods—a flexibility that, as will be discussed later, may be valued by both partners.

The interests of the foreign and local partners, however, are by no means identical. Whereas the local partner can be expected to strive for maximal returns to the joint venture, the foreign partner is more likely to aim for maximal benefits to the whole multinational network, of which the joint venture is only one unit. Because of this difference in objectives, the foreign partner will resist pressure to increase domestic content longer and more vigorously than will the local partner.

The diverging attitudes of the private partners regarding domestic content is mainly due to their different views of the economics of supply. Supplying the local venture with materials from an affiliated source entails profits for the foreign parent. Therefore, in the case of a wholly foreign-owned venture, the foreign parent would be willing to switch to a local source of supply only after the local price falls below the marginal production cost of comparable supply from the affiliated source. In the case of a joint venture, the foreign partner would not want to switch unless its own share of the resulting savings exceeds the profit contribution that it receives by supplying the joint venture from an affiliated source.

The local partner, on the other hand, needs to consider only the *price* charged by the foreign affiliate and the cost of local supply: as soon as the latter drops below the former, the local partner would find it economical to switch to local supply.

The foreign partner may also have other reasons, some not so easily measured, for resisting local supply. If the foreign partner is pursuing a policy of production rationalization, he may consider the joint venture's desire for independent sourcing of parts as disruptive to the rest of the multinational system, even if such a move entails no economic loss for the foreign partner.[5] It is also reasonable to expect the foreign partner to be more concerned than the local partner with the quality image of the joint venture and its reflection on the foreign partner's overseas markets.[6]

The few studies that have addressed this proposition have mostly concentrated on the effect of foreign ownership in the context of a comparison between wholly foreign-owned subsidiaries and locally owned licensees. In a study of Mexican auto firms, Edelberg found that locally owned firms relied on local materials to a larger extent than did foreign-owned firms.[7] Brash, in a study of American investment in Australia, concluded that joint ventures imported a smaller propor-tion of their material requirements than did wholly American-owned firms.[8] He also found that joint ventures depended on their foreign parents for a substantially smaller portion of their imports than did wholly owned subsidiaries. Safarian has obtained similar results for foreign firms in Canada.[9] More recent evidence includes a large study of foreign investment in less developed countries by Reuber[10] and an investigation of foreign firms in Nigeria by Biersteker,[11] both of which reported a significant positive relationship between foreign ownership and import content.

The analyses by Safarian, Brash, and Reuber did not account for the effect of industry differences among their sample firms. Brash himself cautions that the lesser reliance of joint ventures on imports may reflect the fact that a large proportion of joint ventures had originated as fully Australian-owned operations with an established pattern of local purchasing. Commenting on the Brash study, Stopford and Wells also raise the possibility that imports of the wholly owned subsidiaries may have been overstated because the subsidiaries were likely to have been charged more than the joint ventures for their purchases from foreign affiliates.[12] Finally, the statistical significance of the results obtained by Biersteker may have been due more to a pooling of time-series and cross-sectional data than to the underlying pattern observed.

In sum, prior work generally supports the hypothesized relation-ship, although the findings are clouded by the problem of incompara-bility of sample firms.

To explore how foreign ownership and managerial control of the sample firms influenced their local integration (or, equivalently, their import dependence), data were collected for as many years as possible since the start of operations on total manufacturing costs, imported materials, and the proportion of imported materials that consisted of finished component parts ready for assembly.

The proportion of a joint venture's operating life that had been spent under a foreign managing director, appointed by the foreign partner, was used as a proxy for the degree of foreign managerial control over the joint venture, since the managing directors of the sample firms appeared to have a strong influence over decisions most vital to the future of the firm.

A number of analytical tools were used to determine whether import dependence was significantly related to foreign partner ownership share or managerial control. The simplest approach was to form two-by-two contingency tables interrelating different variables.

The results were contrary to expectations: only four of the eleven joint ventures with a low foreign equity participation (0 to 24 percent of total equity) had manufacturing value added of 30 percent or greater, whereas twelve of the seventeen firms with a high foreign equity participation (more than 24 percent of total equity) did so. (Manufacturing value-added was measured by subtracting from unity the ratio of imported materials to total manufacturing costs $\left[1 - \dfrac{IM}{TMC}\right]$.) Likewise, out of thirteen joint ventures that had a foreign managing director during most of their operating lives, only two had value added of less than 30 percent of manufacturing cost, as compared with ten out of fifteen that had been mostly under Iranian managerial control. An analysis of the import figures also suggested a *positive* correlation between foreign ownership/managerial control and the extent of local integration. A test of the effect of the firm's age did not change the outcome.

There was a great deal of interindustry variation in local content statistics, however. It was reasonable to expect that differences in technological complexity would profoundly affect the ease with which firms could manufacture components in Iran. Consequently, matched-pair analysis was used to correct for the effect of industry variations. Nine pairs of firms were selected, with each pair comprising two firms engaged in manufacturing nearly identical or very similar products and utilizing comparable technology, while differing from one another only in terms of foreign ownership, or managerial control, or both.

Five of these nine pairs were manufacturing practically the same product.

In seven of the pairs, the firm with a greater foreign ownership share also had a greater degree of foreign managerial control. In two pairs, however, greater foreign ownership was associated with less foreign managerial control. Therefore, these firms (pairs 8 and 9) were sometimes designated as "high" and sometimes as "low," depending on whether ownership or managerial control was the basis of classification.

With the effect of industry variations filtered out, a strong relationship emerged between foreign ownership and dependence on imports. Table 10–1 shows that the high–foreign-ownership firms as a group imported an average of 88 percent of their total material requirements, as compared with 78 percent for the low–foreign-ownership firms. In other words, the firms with low foreign ownership obtained almost twice as large a share of their materials domestically as did the firms with high foreign ownership. Moreover, 74 percent of the imports of the high–foreign-ownership firms took the form of finished component parts, as compared with 64 percent for the other group. (Under one-tailed t tests, these differences within pairs were significantly greater than zero at the $\alpha = .05$ level.) For every measure listed, there was only one matched pair in which the member firm with a low foreign share was more dependent on imports than the member with a high foreign share, and there were at most two pairs for which foreign ownership share did not make any difference. The results could not be attributed to differences in the length of operating experience; the high–foreign-ownership group had an average of 5.7 years of commercial operations, as compared with an average of 6.0 years for the low–foreign-ownership group.

Foreign managerial control was also associated with greater dependence on imports (see table 10–2), although this association was not as strong as that between foreign ownership and imports. While the high–foreign-control group imported a slightly higher proportion of its total material requirements than the low–foreign-control groups, this difference was not statistically significant. Finished components, however, constituted a substantially higher proportion of the total imports of the high–foreign-control group than was true for the low–foreign-control group: 74 percent versus 63 percent. This difference was statistically significant (at $\alpha = .01$, using the one-tailed t test),

TABLE 10–1

IMPORT DEPENDENCE OF MATCHED PAIRS (OWNERSHIP CLASSIFICATION)

PAIR	(I) IMPORTED MATERIALS / TOTAL MATERIALS — FOREIGN OWNERSHIP			(II) IMPORTED COMPONENTS / IMPORTED MATERIALS — FOREIGN OWNERSHIP			(III = I × II) IMPORTED COMPONENTS / TOTAL MATERIALS — FOREIGN OWNERSHIP		
	High	Low	Difference	High	Low	Difference	High	Low	Difference
1	90%	90%	0%	100%	88%	12%	90%	79%	11%
2	83	77	6	88	88	0	73	68	5
3	82	68	14	93	86	7	76	59	17
4	100	57	43	85	79	6	85	45	40
5	100	100	0	39	16	23	39	16	23
6	100	95	5	100	77	23	100	73	27
7	46	58	-12	70	45	25	32	26	6
8	100	90	10	24	40	-16	24	36	-12
9	90	66	24	65	56	9	59	37	22
Mean	88%	78%	10%	74%	64%	10%	64%	49%	15%
t ratio (α)			1.89 (.05)			2.28 (.05)			3.07 (.01)

TABLE 10-2

IMPORT DEPENDENCE OF MATCHED PAIRS (MANAGERIAL CONTROL CLASSIFICATION)

PAIR	(I) IMPORTED MATERIALS / TOTAL MATERIALS FOREIGN CONTROL			(II) IMPORTED COMPONENTS / IMPORTED MATERIALS FOREIGN CONTROL			(III = I × II) IMPORTED COMPONENTS / TOTAL MATERIALS FOREIGN CONTROL		
	High	Low	Difference	High	Low	Difference	High	Low	Difference
1	90%	90%	0%	100%	88%	12%	90%	79%	11%
2	83	77	6	88	88	0	73	68	5
3	82	68	14	93	86	7	76	59	17
4	100	57	43	83	79	4	85	45	40
5	100	100	0	39	16	23	39	16	23
6	100	95	5	100	77	23	100	73	27
7	46	58	−12	70	45	25	32	26	6
8	90	100	−10	40	24	16	36	24	12
9	66	90	−24	56	65	−9	37	59	−22
Mean	84%	82%	2%	74%	63%	11%	63%	50%	13%
t ratio (α)			0.38 (*)			2.96 (0.01)			2.29 (0.05)

*$\alpha > .05$.

indicating that highly foreign-managed joint ventures had integrated backward to a lesser extent than the locally managed joint ventures.

The difference in the statistical significance of the results associated with the two classifications seems to indicate that the equity share of the foreign partner was a more dependable indicator of import dependence than was managerial control. More importantly, the data suggest that, when the foreign partner had a substantial equity share as well as managerial control, the joint venture was likely to be more heavily dependent on imports than when the foreign partner had a low equity share *or* had conceded managerial control to the local partner.

Why should ownership be a better predictor of import dependence than is managerial control? The difference may be more apparent than real, for two reasons. First, the proxy used for measuring managerial control (i.e., the nationality of the managing director) may have understated the true degree of foreign managerial control in cases where the managing director was an Iranian. In these cases, the foreign partner may have maintained significant influence over the selection of supply sources and local integration through top-level expatriates serving in sensitive posts under an Iranian managing director.

Second, significant control over policy decisions was exercised through the board of directors. It is reasonable to expect the board to have a decisive influence over issues related to expansion of local integration. Compared with the designation of operating management authority, board membership was usually a closer reflection of the ownership pattern; thus board influence may be partly responsible for the greater strength of the association between the ownership variable and local integration.

The statistical results presented so far should be treated cautiously, as the import statistics on which they are based may be biased in at least two ways. First, as discussed later, strong foreign ownership and managerial control were associated with high transfer prices on imported goods, so the import figures overstate the actual imports of the high–foreign-ownership firms relative to those with smaller foreign ownership. Second, the imports of the most locally integrated producers were overstated because deletion allowances did not reflect the true extent to which local production had replaced part of imported component kits.[13] These two types of bias work in opposite directions and thus may be expected to reduce the overall bias to some extent.

More important, in-depth case studies tend to confirm the patterns observed in the statistical analysis. Extensive interviews, however, also revealed that not all foreign partners and managers were dedicated opponents of local integration, nor did all Iranian industrialists and managers embrace every opportunity to expand the scope of their operations and to reduce imports.

Given the government's objective of expanding local integration, it was reasonable to expect strong public-sector ownership and control to be associated with low levels of imports. The data collected for this study provided some support for this hypothesis. Excluding the three privately owned matched pairs in the automotive industry (which were particularly import intensive), the remaining twelve joint ventures were grouped into two sets of six firms each, in such a way that the combined public and semipublic (mostly investment banks partially financed by the government) sector ownership constituted a majority in one group and a minority in the other. The two groups did not differ significantly in terms of their dependence on imports. But when the same twelve joint ventures were categorized according to whether their managing directors had been nominated by the public/semipublic sector or by the private owners (local and foreign), some differences did emerge. The publicly managed firms were significantly less dependent on imports of finished components than were the privately managed firms: imported components were 42 percent of total imports for the former group, as compared with 67 percent for the latter.

Apart from direct control of manufacturing decisions, the foreign parent also appeared to exert some influence through contractual agreements, principally license and technical assistance agreements. A total of thirty-six such agreements, affecting twenty-five of the sample firms, were examined. Five types of clauses affected local sourcing. Delimiting clauses in a few cases explicitly restricted the scope of local manufacture. Tied-purchase clauses, present in 72 percent of the contracts, restricted the sources from which a licensee could buy. Quality-control clauses had similar potential to restrict the sources of supply. Provisions prohibiting the manufacture of products using third-party know-how limited the ability of a licensee to obtain skills to make inputs. Finally, deletion allowances reduced local purchase incentives, since the deletion price was based on marginal cost.

FOREIGN CONTROL AND INDIGENIZATION

The degree of manpower indigenization was used as an indicator of the extent to which know-how had been transferred. Some authors have argued that imported manpower adds to the available store of knowledge within the host economy[14] and that know-how transfer can be measured in terms of the degree of reliance on imported manpower. But the technological and managerial expertise embodied in imported manpower cannot be considered a permanent addition to the host country or firm; after the initial stages of setting up the operation and training local substitutes, continued reliance on expatriates severely curtails the dissemination of valuable skills to the local economy.

Although not directly related to the question of how best to measure the extent of know-how transfer, the cost of imported manpower is also an issue. Ample evidence indicates that foreign managers and technical experts cost far more than local counterparts.[15] Furthermore, a large portion of the foreign expert's total compensation is likely to be remitted abroad, spent on imported goods, or otherwise removed from circulation in the local economy.[16]

Closely related to the cost question is the sensitive issue of the quality and effectiveness of imported manpower in developing countries. A few studies, including the present one, have noted widespread dissatisfaction on the part of local government authorities and businessmen, as well as multinational companies themselves, with the way in which expatriate personnel have discharged their responsibilities.[17]

Finally, indigenization is clearly tied to the issue of control at the topmost levels of the organization. One might expect the host country's own nationals to be more sensitive to its interests than a foreign manager, whose ultimate loyalty is to an exogenous employer. Even if local managers do not take it upon themselves to protect national economic interests, they are probably more amenable than foreign managers to various means of persuasion adopted by the government.[18] In a joint venture, the private interests of local managers are closely tied with those of the local partner. Just as often, the expatriate will direct his loyalty to the foreign parent organization, where his long-term career goals are most likely to reside.

But some studies suggest otherwise.[19] And the studies of technology choice in this book (Chapters 2 through 7) suggest little difference in behavior between foreign and local managers, including local managers of firms owned by the local government. Yeoman's study is an exception; he found that subsidiaries of foreign firms under local management are more likely to adapt their technologies to local needs and conditions than are those under foreign managers, but his evidence is relatively weak.[20]

There are a variety of reasons to expect the foreign partner of a joint venture to favor a greater reliance—and for a longer period of time—on imported managerial and technical manpower than would the local partner. First and foremost appears to be the issue of control. Given the loyalty of expatriates to the foreign partner organization, and given the multitude of the issues in which the interests of the local parties and the foreign partner are clearly at odds, it is only natural to expect the foreign partner to insist on occupying the key positions in the joint operations.[21] In the words of one researcher:

> ... there may be some inclination to keep a person from the international firm as head of the subsidiary and perhaps in one or two other key posts, well after the early years of the establishment of the subsidiary, to insure that the interests of the international firm are kept to the foremost as the subsidiary settles more fully into the national environment.[22]

Brash maintains that foreign firms favor greater reliance on imported manpower because it facilitates communication between the two corporate entities, and because of a belief that expatriate personnel have a "better attitude" toward business than do nationals.[23] Hymer explains this bias in favor of expatriate personnel in terms of a need for cultural affinity at the upper levels of the organization.[24]

In some cases, the foreign partner may assign expatriates to the local operations as a means of training them for subsequent positions in the headquarters or in other subsidiaries.[25] In other cases, training and supervision of nationals may be regarded as an expensive and unnecessary distraction from the expatriate staff's operating responsibilities.[26] It may seem much quicker, and therefore more advisable, to send an expatriate than to wait until a local national has been trained in the ways of the parent company.[27]

That the local partner would be less disposed than the foreign partner toward employment of imported manpower is also indicated by an analysis of cost considerations. In many circumstances, the ultimate cost to the foreign partner of finding, training, and promoting a national to replace an expatriate may far exceed the benefit gained by recalling that expatriate home. Although the expatriate's salary and expenses are high, they are paid by the joint venture and are thus only partially borne by the foreign partner. Indeed, in some cases the opportunity cost of the expatriate to the foreign parent may have been near zero because of slack in the worldwide organization.[28] It is sometimes charged that multinational firms attempt to unload their superfluous personnel on their foreign operations.[29] The local partner, in contrast, is likely to find that his share of total expatriate expenses exceeds the cost of a comparable local substitute, and thus he will favor a transition as soon as a local substitute is trained.

The foreign partner's bias in favor of imported manpower may also reflect purely technical considerations. More concerned than the local partner with the firm's quality image and brand name, the foreign partner may insist on filling key technical positions such as quality control. Greater reliance on imported manpower is also sometimes defended on the grounds of efficiency,[30] although it is not certain that expatriate personnel are invariably superior to local alternatives in this respect.

In summary, the foreign partner in a joint venture is likely to be more disposed than a local partner to assign high-level positions to expatriates. Thus one would expect a high degree of foreign ownership and control to be associated with heavy reliance on imported manpower.

Few studies have actually tried to test the validity of this assertion. Biersteker found that foreign firms operating in Nigeria relied on imported manpower to a greater extent than did indigenous firms.[31] The Reuber study, involving foreign firms in different developing countries, also found a positive but not highly significant relationship between foreign ownership share and the employment of expatriates.[32] A study by the Reserve Bank of India indicated that wholly foreign-owned subsidiaries and joint ventures in that country utilized expatriates to a greater extent than did licensing operations,[33] and Balasubramanyam corroborated this finding in his interview study.[34]

In this study, two expatriate employment measures were used for analysis: the number of expatriates as a percentage of salaried staff, and the number of expatriates as a percentage of total employment. Unfortunately, employment data for most firms were such that it was not possible to distinguish between professional and nonprofessional staff. The category of staff used in the first measure comprises all salaried personnel, including secretaries and clerks.

Almost all of the expatriates occupied top managerial and technical positions. In most firms, even those with Iranian managing directors, almost all key managerial and technical decisions were affected directly or indirectly by the expatriate staff. This influence appeared to be even stronger for decisions whose consequences extended beyond the national borders, such as those affecting foreign purchasing and relations with the foreign partner.

Although the absolute number of expatriates employed by the sample firms was not inordinately large, maintaining them in Iran often constituted a significant financial burden. It was not unusual for overseas allowances, living expenses, and payments to the parent firm to amount to a premium of 200 percent over the expatriate's base salary. In one firm the total expatriate compensation stood at 2.5 percent of the cost of sales, roughly equal to the direct labor cost of production. In another venture, total expenses per foreign manager averaged just under $10,000 per month, equivalent to 8 percent of sales at a time when the company was reporting losses. The total selling and administrative expenses of a third joint venture declined by more than 70 percent after the services of its foreign management contingent were terminated.

Joint ventures with a high degree of foreign ownership relied on imported manpower to a greater extent than did those with a low foreign share (see table 10–3). The two groups of firms did not differ significantly in terms of their age, which averaged 6.3 years for the low–foreign-ownership group and 6.1 years for high–foreign-ownership group.

Dependence on expatriate manpower also appeared to be positively associated with the degree of managerial control by the foreign partner. (As before, nationality of the managing director was used as the proxy for the locus of managerial control. To avoid a tautology, foreign managing directors were not included in the total expatriate count.) While only four of the seventeen firms with an Iranian managing director had an expatriate-to-staff ratio in excess of 5 percent, this was the case with more than half of those with a non-Iranian managing director. Again, the small difference in the average age of the two

TABLE 10–3

FOREIGN OWNERSHIP AND DEPENDENCE ON EXPATRIATES IN 28 COMPANIES

| | | FOREIGN OWNERSHIP SHARE OF COMPANY | | |
		0–24%	≥25%	Total
Expatriates as a percentage of salaried staff				
	<5%	10	7	17
	≥5%	1	10	11
Expatriates as a percentage of total employment				
	<1%	10	6	16
	≥1%	1	11	12

NOTE: Figures indicate number of firms.

groups was unlikely to account for much of the difference in dependence on expatriate personnel.

As a further check, matched-pair analysis was used to explore the possibility that the observed differences in dependence on expatriates were due to differences in technological complexity. Nine pairs of firms were again selected and, as before, the analysis was done separately on two sets of "high" and "low" foreign influence classifications: one based on the ownership share and the other based on managerial control (i.e., nationality of the managing director).

The results of the matched-pair analysis confirmed the conclusions obtained so far. For the high–foreign-ownership firms, expatriates made up 10.1 percent of salaried staff, on average, as compared with 3.2 percent for the low–foreign-ownership group. Average expatriate-to-total employment shares were 1.8 percent and 0.5 percent, respectively. (The differences in the expatriate-to-staff ratios were statistically significant at the $\alpha = .02$ level; and in the expatriate-to-total employment ratio, at the $\alpha = .01$ level.)

Historical data available for seven of the nine matched pairs made it possible also to compare the speed with which the two groups had reduced their dependence on imported manpower. For these seven pairs, the high–foreign-ownership firms as a group reduced their actual number of expatriates from an average of 7.4 to an average of 5.2 in a little over four years. The firms with a low–foreign-ownership share reduced their dependence more drastically, from an average of 6.7 to 2.8, in just under five years.

When the basis for the "high" and "low" classification was changed from ownership to managerial control, the same conclusions were supported even more strongly. Furthermore, none of the observed differences could be explained by either the size or the age of the firms.

Information gathered through interviews generally supported the foregoing reasoning and evidence. The predominant impression that emerged from these interviews was of the gap between the views of the Iranian and the non-Iranian managers. Iranian managers and industrialists generally tended to believe that their foreign partners put too little emphasis on executive development, insisting instead on relying on personnel supplied from abroad. This feeling appeared to be particularly intense in the middle-management ranks.

Expatriates, on the other hand, generally viewed their presence as indispensable to the well-being and indeed the survival of the operation. Some of them felt that expatriate manning levels should ideally be even higher than they were in practice.

Not all Iranian managers and businessmen were opposed to the employment of expatriates. In particular, overcommitted businessmen who were involved simultaneously in a multitude of manufacturing and trading projects found it convenient, if not imperative, to entrust management of the joint venture to the foreign partner, regardless of the cost. Similarly, the employment of expatriates was quite acceptable to those local shareholders who owed their equity participation to their political position or influence rather than to an actual contribution of capital by them or to their possession of valuable business expertise.

Reluctance to rely on expatriates also varied in other ways. Most of the opposition to the employment of expatriates that was voiced by Iranian managers and industrialists centered on managerial positions; much less concern about purely technical jobs was evidenced. An expatriate working in a technical capacity was not perceived to be in a position to infringe on the local partner's control of the venture, at least not as much as one in a managerial position. Moreover, many Iranians believed that expatriates did not perform as well in managerial positions as in purely technical ones.

Some Iranian managers expressed a preference for the short-term employment of expatriates hired for specific tasks such as training or solving technical problems as they arose. In those situations, it was not necessary to make long-term residence arrangements (moving the family, schooling for the children, etc.), and there were fewer adjustment problems for the expatriate himself. Moreover, when the expatriate was hired for a specific task, his authority and responsibility could be more sharply defined, thus reducing the potential for conflict of interest with the joint venture.

Some managers, however, doubted the advisability of relying heavily on short-term expatriates, arguing that a warm-up period was necessary before they could be expected to reach their best performance. And satisfactory short-term manpower might not always be available in sufficient quantity and upon short notice. In any case, reliance on short-term manpower appeared to be favored by firms that were well advanced in their development and thus could determine the exact nature of their manpower requirements independently of the foreign parent, and so could supervise such staff effectively.

Feelings about expatriate employment also depended on whether line or staff responsibility was involved. A few local managers preferred to assign only training and advisory roles to expatriates. One Iranian manager stated that he was reluctant to give line responsibility to an expatriate who normally could not be evaluated and compensated on the basis of actual performance.

Finally, a few Iranian managers raised the possibility of directly hiring expatriates not affiliated with the foreign parent. This approach would largely resolve the question of loyalty, although lack of familiarity with the foreign parent organization would be a disadvantage. Reliance on nonaffiliated expatriates, in any case, appeared feasible only after the joint venture had acquired substantial operating experience and had become capable of clearly identifying its technical manpower needs. Proper recruitment, evaluation, and selection of nonaffiliated expatriates appeared to require more resources than were available to many of the younger operations.

FOREIGN CONTROL AND THE COST OF TECHNOLOGY

Foreign firms can use several methods to charge their affiliates for the transferred technology and know-how. Royalties, technical assistance fees, and various service charges are the explicit ways in which payments are received. When the parent firm supplies raw materials, component parts, and machinery to the affiliate in addition to know-how, or when it purchases finished goods from the affiliate, the transfer prices used in these transactions can provide an indirect means of extracting an income from the affiliate.

When the affiliate is a joint venture, the parent company's motivation to charge high transfer prices is particularly strong. The parent firm has the incentive to take as much of the economic rent out of the joint venture as possible, rather than share it with the local partner through profits and dividends. In fact, relatively little is known about the extent to which, and under what conditions, multinational enterprises charge higher-than-market prices to maximize their profits.[35] But to the extent they desire to do so, the presence of a local partner is likely to constrain the parent company's freedom to fix transfer prices arbitrarily. To protect his own interests, the local partner is motivated to search for alternative sources of supply and to challenge the prices charged by the foreign parent if they are higher than those quoted by independent suppliers. Thus as the foreign parent company's ownership share declines, its incentive to overcharge the joint venture for material inputs increases, but its ability to impose excessive prices can be expected to decline. Consequently, locally managed joint ventures are likely to pay lower prices for their purchased materials than joint ventures managed by the parent company.

The potential for overpricing—that is, the extent of host-country dependence on material imports—appears to vary substantially, depending on the countries and industries involved. For example, Mason maintains that the potential for U.S. multinational firms to repatriate profits from their Latin American operations via transfer pricing was limited because, as early as 1966, imports represented only about 8 percent of the value of the output of these operations, and no more than 10 percent of the total output was exported.[36] In the same vein, Chudson and Wells report that only about 15 percent of the exports of U.S. multinational firms in 1966 were intermediate goods shipped to their affiliates in the developed and developing countries, although this proportion was likely to be much higher for affiliates located in the developing countries.[37]

Streeten and Lall, on the other hand, maintain that the volume of intermediate product imports into many developing countries is sufficiently high to allow substantial transfer of funds through indiscernibly low levels of overpricing. Other studies have produced some evidence in support of this contention. For example, Vaitsos reports that foreign affiliates accounted for 48 percent of Colombia's total imported intermediate products in several principal industries, and that over one-half of the exports of U.S. subsidiaries in that country were to affiliates around the world. He found that imported materials constituted between one-half and three-quarters of the value of output in several firms in the Colombian pharmaceutical, chemical, rubber, and electronics industries.[38]

Information regarding actual transfer prices has, understandably, been difficult to obtain. Yet some evidence on overpricing has begun to accumulate, especially in the pharmaceutical industry. Vaitsos found that foreign subsidiaries in the Colombian pharmaceutical industry were charged for imported intermediate products at prices that were on the average 155 percent more than prices from the lowest-cost sources. The funds removed from these subsidiaries through overpricing amounted to six times the royalties and twenty-four times the declared profits. Similar but less dramatic findings were obtained for the same industry in Peru and Chile, and for the chemical, rubber, and electronics industries in Colombia. The extent of overpricing experienced by wholly foreign-owned subsidiaries in all three countries was consistently and significantly higher than that experienced by locally owned firms.[39]

A study of the pharmaceutical industry in Iran by Salehkhou uncovered substantial evidence of overpricing of imported materials, ranging up to more than ten times the free-market prices.[40] Surveys conducted by the United Nations Conference on Trade and Development (UNCTAD) in Spain, Ethiopia, Mexico, and Ecuador provided additional evidence of overpricing.[41]

Mounting evidence indicates a positive relationship between foreign ownership share and the extent of interaffiliate purchases. The Brash and Reuber studies both found that wholly foreign-owned subsidiaries depended on their foreign parents for a greater portion of their imports than did joint ventures.[42] The evidence regarding transfer pricing, however, was mixed. On the basis of interview responses, Brash concluded that, while some firms in Australia were being overcharged, others paid "fair" prices and still other firms reported paying below-market prices for inputs purchased from parents. His data suggest a nonlinear relationship between foreign ownership share and overpricing: the great majority of firms being either overcharged or undercharged were wholly owned subsidiaries, whereas the majority of firms being charged arm's-length prices were joint ventures. Robbins and Stobaugh, in their study of the financial practices of U.S.-based multinational enterprises, found it difficult to draw unequivocal conclusions about practices on transfer pricing. But they did conclude that firms do not by any means exploit the full potential of transfer pricing as a deliberate means of shifting funds from one entity to another.[43]

The present study found the potential for overpricing to be very high for the majority of the firms. Imports of raw materials and component parts constituted a large fraction of the total manufacturing costs of the twenty-six sample firms for which data were available. Imported materials accounted for 30 percent to 86 percent of manufacturing costs (defined as sales less profits, interest, and all administrative overheads), with an unweighted average of 59 percent. With the cost of imported machinery taken into account, import dependence was even greater. The sum of capital equipment depreciation charges and the cost of imported materials varied between 38 percent and 90 percent of total manufacturing costs, with an unweighted average of 67 percent.

Estimates provided by sixteen joint ventures clearly indicated the dominance of the foreign parent corporation as the source of imports. About half of these firms purchased their entire import requirements from, or through, their parent corporations; only two firms reported not purchasing any materials from the foreign parent. On the average, 73 percent of the imports of these firms were provided by affiliated sources. More important, there was much evidence of a positive relationship between the degree of foreign control and the extent of reliance on the foreign parent as the main source of imports.

The regulatory environment in Iran had eliminated some of the incentives for overpricing. Repatriation of profits and capital was not restricted, nor were interaffiliate charges tightly controlled. Foreign exchange was plentiful and widely available. Numerous tax holidays removed taxes as an important consideration. At the time of this research (1977–78), there was also a positive attitude towards business in general, and the country had enjoyed a long period of political and economic stability.

Still, the price control and protection policies of the government provided some incentive for overstating the cost of imported materials. Tariffs on imported raw materials, components, and capital equipment were too low to deter this tendency.

About one-third of the sample firms provided at least some direct evidence of discriminatory transfer pricing. The best-documented case shared with this researcher involved a firm that assembled a certain piece of heavy equipment from over 400 component parts. Initially managed by a team of expatriates, this 30 percent foreign-owned joint venture had relied on the foreign parent as the sole supplier. Following local takeover of management, however, the license agreement had been renegotiated, allowing the joint venture direct sourcing from a list of approved suppliers. Taking advantage of this freedom, the joint venture had aggressively solicited new suppliers, significantly reducing its dependence on the foreign parent and realizing substantial savings. A lengthy list compiled by the supply manager indicated savings on several component parts ranging between 17 percent and 60 percent of the prices that had been charged by the parent firm. He estimated that these savings had reduced the import bill by more than 15 percent in a single year. This saving was very significant, because imported materials in that year represented more than 85 percent of the total manufacturing cost while direct labor costs amounted to no more than 2.3 percent. That is, estimated savings in material costs had been more than five times the production-wage bill.

Another joint venture, 50 percent foreign owned and fully foreign managed, had signed a supply agreement obligating itself to purchase its entire requirements of a certain high-volume raw material from the foreign parent. Within five years after start-up, the foreign parent had almost quintupled its sales price to the joint venture, while the free-market price had barely doubled. As a result, the cost of raw materials had risen from 27 percent to 57 percent of total manufacturing costs, while total labor costs had remained steady at around 16 percent. Saddled with significantly higher raw material costs, the joint venture had started to lose market share to imports and to incur substantial losses. In an application to the government for higher tariff protection (to be effected by raising the tariff on the finished product), the company had stated that in just one year the price of the raw material had doubled whereas the import price of the finished product had declined by about 40 percent. The higher tariff protection was granted and immediately passed on to customers in the form of higher product prices. The supply contract eventually was amended at the insistence of the local partners, allowing the joint venture to purchase part of its requirements from other sources. Interestingly, the foreign parent began quoting lower raw material prices soon after the contract was renegotiated.

Other isolated cases of overpricing were clearly identified. One foreign firm had sold used equipment to its local venture and charged the price of new equipment. Upon discovery by the local partner, the foreign parent had agreed to pay back the difference between the price actually charged and the assessed value of the equipment. The Iranian managing director of another firm reported having to purchase component kits from its foreign parent at 40 to 80 percent above the overseas price of the product in assembled form. Two joint ventures reported being charged extremely high prices for spare parts, ranging up to six times the prices quoted by the original manufacturer of the equipment.

Other joint ventures complained about excessive commissions charged by the parent for handling purchases from independent suppliers. Commissions reportedly charged were 88 to 149 percent of the original purchase price in one case, 100 percent in another, and 21 to 67 percent in a third. One Iranian managing director stated that the parent company had rejected his offer of a flat commission of 20 percent. He estimated that, given large-volume purchase discounts normally received by the parent, it probably enjoyed a 50 percent profit margin on the purchases handled for the joint venture.

Indirect evidence of overpricing was also noted. One majority–foreign-owned pharmaceutical firm that imported its entire raw material requirements from its parent company had reported losses over five years amounting to a little less than half of its share capital. Over the same period, a wholly Iranian-owned pharmaceutical firm with lower sales volume had reported an average after-tax return of 31 percent on sales and 39 percent on equity. Commenting on the majority–foreign-owned firm, one industry analyst observed:

> ... they are either very inefficient in spite of their large contingent of foreign experts, or profits are being siphoned off .... But they are very careful not to lose more than half of their share capital because they would then have to declare bankruptcy. ... The minority Iranian partner wants to sell his shares but

there are no buyers. The parent company will probably end up
buying them which is something they have always wanted to do.

Curiously, despite continuing reported losses, this firm had twice
expanded its output capacity.

A similar situation was observed in the automotive industry. One
minority–foreign-owned assembler had consistently reported low
profits or losses in the booming automotive market while its major
direct competitor, a locally managed licensee, had reported consistently high profits. Another foreign-managed assembler with the
largest foreign equity share in the industry had been charged for
imported knocked-down kits at prices that were only marginally below
the retail price of the assembled car (including factory profits, excise
taxes, and dealer margin) in the European market. This firm had also
reported losses for two years in spite of heavy protection and high local
prices and yet had extensive plans for expansion.

Most of the evidence on excessive transfer prices presented above
related to joint ventures in which the foreign partner had a large equity
share and significant managerial control. Overpricing was infrequently
reported by locally managed joint ventures, and transfer prices were
often reduced after local takeover of management, suggesting that
local control was successful, at least to some extent, in countering
excessive input prices. The evidence is limited, however, and allows
only a tentative conclusion on this point.

Another category of technology costs involves direct and explicit
payments by the joint venture to the foreign parent corporation:
royalties paid for transfer of patents, trademarks, and production
technology; and service charges such as management fees, engineering
fees, training charges, and other miscellaneous charges, which are paid
in return for specific services rendered by the foreign parent.

Some authors have argued that wholly or substantially owned
operations are less likely to be charged by their parent firms for know-
how than are licensees or firms in which the foreign equity share is
small. Since all or most of the economic returns from wholly or
substantially owned operations accrue to the foreign parent anyway,
there may be less need to charge for the resources transferred.[44]

To explore the effect of foreign ownership on the magnitude of
royalties, it was first necessary to translate such payments into a
uniform measure for all firms—as a percentage of sales and of manu-

facturing costs. This was possible for twenty-eight firms. Measured in this way, royalty payments did not seem very large: they ran only an average of 1.8 percent of sales and 2.3 percent of manufacturing costs. Royalty payments averaged 7.5 percent of total manufacturing value-added, however, reaching 33.1 percent and 21.7 percent for two firms, both of which were majority foreign owned. Fully one-fourth of these twenty-eight firms paid more than 10 percent of their value-added in royalties. Over the whole sample, royalty payments averaged 20.8 percent of manufacturing wages and salaries. (Manufacturing wages and salaries included the salaries and expenses of expatriates employed in the plant, which often constituted a significant portion of the total wage bill. Their exclusion would have further magnified the importance of royalties.) For one pharmaceutical firm, this rate had averaged 70.3 percent over the five-year period for which data were available, reaching as high as 135 percent in a single year.

The parent companies' royalty incomes appeared to be small in comparison with their earnings from sales of materials to the joint ventures. To get an idea of the magnitudes involved, royalties were calculated as a percentage of imported materials. Of twenty-two firms for which this ratio could be computed, the royalty payments were below 5 percent of the value of the imported materials in fourteen cases. For only two firms (which did not import a great deal) was the ratio above 10 percent. The average ratio for the twenty-two firms as a whole was slightly over 4 percent. Given the evidence on transfer prices, and considering the fact that sales of materials usually involve a considerable profit margin even without overpricing, it is clear that sales of inputs to these joint ventures were typically a much more significant source of income to the parent than were know-how charges.

Analysis of royalty figures revealed a significant *positive* relationship between foreign ownership and the magnitude of royalty payments. Royalty payments as a percentage of sales were strongly correlated with foreign equity share of the twenty-eight firms ($R^2 = .52$ and a positive β coefficient that was significantly different from zero [$\alpha < .01$]). Statistical tests using the other royalty measures as dependent variables resulted in a similar conclusion.

Even when the effect of industry variations was removed by focusing on matched pairs, the conclusion remained basically unchanged, although its statistical significance was reduced. For seven of the nine matched pairs the member firm with the higher foreign-ownership share paid at least as much in royalties as the corresponding firm with the lower foreign share. The highly foreign-owned firms as a group paid an average of 2.1 percent of their sales in royalties, as compared with an average of 1.6 percent for the firms in which foreign equity share was small. These differences in royalty payments, however, were not highly significant ($\alpha \cong .15$).

The conclusion that the magnitude of royalty payments was positively correlated with the degree of foreign ownership runs counter to the arguments advanced earlier and against most of the existing evidence. An interview study of several U.S. multinational enterprises in the chemical industry by Stopford and Wells[45] and the Brash study of American investment in Australia[46] concluded that joint ventures were likely to pay higher royalties than wholly owned subsidiaries. Reuber also found a statistically significant and negative relationship between the extent of foreign ownership and royalty payments.[47]

One explanation for the apparent contradiction between the present study and the others may be that none of the firms studied here was a wholly owned subsidiary. Full ownership may indeed be a necessary condition for the foreign parent firm to refrain from charging royalties. If so, the implication would be that a foreign parent will attempt to charge a joint venture the maximum royalty possible, but that substantial local participation will act as a restraint on the amount actually charged. This explanation appears to be in line with the results of a study by Streeten and Lall that included sixteen joint ventures in Iran.[48] Royalty payments of the majority–foreign-owned firms in their sample were, on the average, over 400 percent of their declared profits, as compared with an average of 60 percent for the minority–foreign-owned firms.

The generally higher royalty payments of the highly foreign-owned joint ventures did not appear to have been offset by lower service charges or capitalized know-how. None of the firms in the matched-pair sample reported paying for know-how in equity shares. Furthermore, management and engineering fees that were available from a number of the firms were insignificant in comparison with royalties and did not vary consistently with the degree of foreign ownership.

FOREIGN CONTROL AND EFFICIENCY

In spite of frequent complaints about the performance of expatriate managers, it is possible that foreign ownership and management control lead to a more efficient use of resources and a more highly valued product than do local ownership and control. If so, the cost implicit in the less efficient operations of locally controlled joint ventures should be balanced against their lower royalty and transfer-price payments and the higher benefits implicit in their greater local value added and indigenization.

Measuring relative efficiency is a difficult task; few studies have gone beyond citing circumstantial evidence, and those studies that have attempted a systematic analysis of differences in efficiency have mostly used highly imperfect measures. For example, in two separate studies of firms in Australia and Canada, Brash and Safarian have used total value of production per employee and value added per employee, respectively, as indicators of productivity.[49] Clearly, differences among firms in terms of labor intensity, the degree of vertical integration, and the degree of protection can introduce serious distortions in such measures. Nonetheless, both studies found that wholly foreign-owned subsidiaries were significantly more productive than locally owned firms. Brash found indications that wholly foreign-owned firms were not generally more productive than jointly owned firms, however.[50]

In an economy where market forces are given a reasonable chance to operate, the question of efficiency can be largely resolved through an examination of simple business performance measures such as profitability, market share, and growth. In the highly regulated Iranian economy, however, competition was restricted in many ways, and the costs and benefits of these restrictions were unevenly distributed. In such conditions, simple business measures cannot provide a valid basis for comparing efficiency.

By making some adjustments in the available data, it was possible to arrive at some reasonable measures and tentative conclusions. Two measures of efficiency were used: degree of competitiveness with imports, and domestic-resource cost of production. Analysis of these measures was carried out on the matched pairs. To focus specifically on the effect of differences in managerial efficiency, firms were classified as "high" or "low" according to the degree of foreign managerial control.

The extent to which a firm can compete in international markets, in terms of the price and quality of its products, is perhaps the best indicator of its efficiency. Most developing countries are not only interested in replacing imports with low-cost local production but also hope eventually to become exporters of manufactured goods. The competitiveness of local firms in international markets is therefore crucial.

Unfortunately, none of the firms in this study had a consistent export record that could provide a basis for judging export performance. Only a few firms had sporadically exported a small fraction of their output, and then mainly as a part of government barter agreements with other countries.

It was possible, however, to compare the sample firms on the basis of their ability to compete with imports. Table 10–4 shows the ex-

TABLE 10–4

LOCAL PRICES COMPARED TO COST OF IMPORT

$$\frac{\text{Local Ex-Factory Sales Price}}{\text{CIF Cost of Comparable Import}}$$

| | FOREIGN MANAGERIAL CONTROL | |
PAIR	High	Low
1	1.39	1.56
2	1.71	1.50
3	1.66	1.37
4	1.03	1.00
5	1.27	2.10
6	1.00	1.73
7	1.12	1.62
8	1.20	0.96
9	1.53	1.76
Mean	1.32	1.51
t ratio	1.18	
(α)	(0.15)	

factory sales price (excluding excise taxes, if any) of the products of the nine matched pairs, as a multiple of the duty-free CIF cost of importing the same product or another brand that the firm considered to be closest to its own. (In cases where different models of the same product were being manufactured, the ratio appearing in table 10–4 either is an unweighted average of the ratios corresponding to different models, or it corresponds to the highest-volume model if complete data were not available.) The table suggests that foreign managerial control was associated with slightly lower local prices. As a group, the nine joint ventures under high foreign managerial control charged an average premium of 32 percent over the cost of comparable imports, as compared with an average premium of 51 percent for the low–foreign-control group. These differences in average premiums, however, were not highly significant ($\alpha \cong .15$).

Local prices clearly included an element of profits. Profits were distorted by price controls and unequal degrees of protection from imports. To see if this distortion had biased the results, unit production costs—rather than sales prices—were compared with the CIF cost of imports. This comparison, presented in table 10–5, did not change the basic conclusion.

The differences in performance disappear altogether, however, when attention is focused on the five perfectly matched pairs; that is, pairs 1 through 4 and 9. (These five joint ventures were producing the same product and therefore were subject to the same rate of protection and level of price controls.) The unit manufacturing costs of both the foreign-controlled and the locally controlled joint ventures in these five pairs averaged about 24 percent higher than the cost of comparable imports.

On the basis of this analysis, therefore, it cannot be confidently concluded that a high degree of foreign managerial control led to significantly lower prices or production costs.

One more issue, however, remains to be resolved. Since the foreign-managed joint ventures generally paid higher prices for their imported materials than did the locally managed joint ventures, the total costs or prices of foreign-managed joint ventures would have been lower if they had paid the same prices for imported materials as did the locally managed ventures. On the other hand, since the foreign-managed firms purchased a lesser portion of their material requirements from local sources than did their locally managed coun-

TABLE 10–5

LOCAL MANUFACTURING COSTS COMPARED TO
COST OF IMPORT

$$\frac{\text{Local Unit Manufacturing Cost}}{\text{CIF Cost of Comparable Import}}$$

| PAIR | FOREIGN MANAGERIAL CONTROL | |
	High	Low
1	1.20	1.26
2	1.42	1.22
3	1.56	1.23
4	0.77	0.90
5	1.00	1.64
6	0.74	1.45
7	0.82	0.94
8	1.00	0.80
9	1.19	1.60
Mean	1.08	1.23
t ratio	1.14	
(α)	(>.15)	

terparts, and since the local materials in some instances may have cost more than imports, the total costs or prices of the foreign-owned ventures would have been higher if they had purchased the same portion of material requirements locally as did the locally managed firms. The information available on transfer prices and the cost of local materials was not sufficient to allow any adjustment of the figures for this possibility.

The second approach to measuring efficiency, which is derived from a method used to analyze projects from the national point of view, does not suffer from the same problems of bias. Economists measure the efficiency of import-substituting enterprises in terms of the opportunity cost of the domestic resources these enterprises use to generate one unit of value added at international prices.[51] One such measure, the domestic resource cost, is the ratio of domestic value added over value added at international prices, and it reflects the cost to the economy of producing a good locally rather than importing it.

Since the intent in this study was to measure the efficiency of the value-adding process within the firm, several adjustments to the normal definition of the domestic resource cost measure were necessary. First, the cost of domestically purchased materials was excluded from the domestic value added measure. Second, since efficient managers could be expected to attempt to optimize their mix of labor and capital on the basis of the market prices of these factors, the actual costs of labor and capital—rather than their opportunity costs to the economy—were used. Third, an adjustment in profits was necessary. With uniform output prices in an industry, a firm's profitability will be inversely related to its costs: the lower these costs, the greater the profitability. Since domestic value added normally includes profits, the domestic resource costs of the most efficient firm would be overstated in comparison with those of other firms in the same industry to the extent of their differences in profits. To alleviate this problem, the actual profit figures were replaced by a uniform rate of return on the equity capital employed.

With these adjustments, value added by each firm was calculated by first deducting from its total costs 1) the landed cost of imported raw materials and components (including duties), and 2) the actual cost of locally purchased materials. Then 20 percent of the total equity base (share capital plus retained earnings) was added to account for the cost of equity. (The results were not very sensitive to the assumed cost of equity.)

Value added at international prices was defined as the CIF value of local sales, less the CIF cost of imported raw materials and components, less the equivalent CIF value of locally purchased materials. The CIF value of sales was derived for each firm by multiplying its ex-factory sales by the ratio of unit CIF price of the product over the ex-factory unit sales price. A uniform tariff rate of 10 percent was assumed for imported components. The results, it turned out, were not sensitive to changes in this assumption. It was also assumed that local prices of materials were 10 percent higher than their CIF value. Again, the results were not sensitive to reasonable changes in this assumption.

The adjusted domestic resource cost measures of the nine matched pairs, estimated in accordance with the above assumptions, appear in table 10–6. The smaller the value of the measure, the more efficient the firm with which it is associated. Foreign managerial control was mildly associated with lower domestic resource costs.

Because of the incidence of negative ratios, it was not possible to determine the statistical significance of the differences in domestic resource costs between members of the two control groups. It should be noted, however, that the two most inefficient firms (i.e., those with negative international value added) were foreign managed. Furthermore, among the five perfectly matched pairs (1 through 4 and 9), only three of the rankings favored the foreign-managed firms.

A final point should be emphasized. The lower domestic resource costs associated with several of the foreign-managed firms may not be due simply to differences in managerial efficiency. Part of the observed higher costs of the group with less foreign control may be attributed to the fact that they had integrated backwards into operations that impose scale problems because of the capital-intensive processes

TABLE 10–6

ESTIMATED DOMESTIC RESOURCE COSTS OF MATCHED PAIRS[a]

Domestic Cost of Value Added
──────────────────────────────
International Value of Value Added

| PAIR | FOREIGN MANAGEMENT CONTROL | |
	High	Low
1	2.3	6.6
2	negative	3.0
3	negative	4.7
4	0.7	1.1
5	1.6	8.7
6	0.7	3.8
7	1.2	1.5
8	1.6	1.0
9	2.1	6.6

[a]Both value-added measures are expressed in local currency with conversions at the official rate.

involved. As a result, some of their machinery was likely to have been less fully utilized than that of the less integrated firms. Underutilization of capital equipment represents a real cost to the firm and the economy, but it may be the result of the choice of industry segments into which the firms integrated, rather than the consequence of inefficient management.

The absence of clear evidence indicating greater efficiency on the part of foreign-managed firms is contrary to expectations. Among the many plausible explanations, one is of particular interest. It is possible that many of the firms in the sample, and indeed in the entire Iranian economy, simply had little motivation to be cost effective. The decade of the 1970s, from which the data for this study were derived, was characterized by sellers' markets in which most products could be marketed easily, almost regardless of price and quality. Furthermore, the regulatory environment and the market structure bestowed near monopolistic positions on many firms. Perhaps most important, widespread corruption and ineptitude of government policy makers and administrators may have driven many managers, regardless of their nationality, to seek profits by bargaining for special deals rather than by increasing productivity. This phenomenon, dubbed "satisficing," has previously been reported in other research.[52]

CONCLUSIONS

The divergent objectives pursued by joint-venture partners lead to conflicts in the division of costs and rewards. The manner in which these conflicts are resolved, and therefore the ultimate distribution of costs and rewards, depend on the relative control of each partner over joint-venture policies and operations. Seeking to maximize earnings for its multinational network as a whole, the foreign partner generally tries to limit the extent of local manufacture and indigenization of management and strives to charge the joint venture the highest possible prices for the goods, know-how, and services supplied. On the other hand, there was evidence, albeit weak, that foreign-partner involvement helped promote and maintain the efficiency and competitiveness of the venture.

This and other studies indicate that adherence to a rigid policy with respect to foreign ownership and control may not be in the best interests of the host country and local private parties. While restricting the degree of foreign ownership and managerial control of joint ventures serves the host country's interests in several ways (higher local integration, greater indigenization of management, and lower transfer prices, royalties, and fees), there may be other costs involved in such a restriction: low efficiency and high production costs.

While real conflicts exist between the interests of the local and the foreign partners, the greater efficiency associated with a high degree of foreign involvement suggests that the local partners should not view the joint venture as a zero-sum game. Rather, the local partner may have to choose between a smaller share in a more efficient (and therefore more profitable) operation and a larger share of a less efficient operation.

A policy that overemphasizes restriction of foreign ownership and control may have other costs as well. The joint venture's access to the foreign parent's export markets and international distribution network may be a function of the parent firm's equity stake. Although none of the firms in this study had engaged in significant export activities, other research has produced strong evidence that foreign firms are more likely to allocate export opportunities to affiliates in which they have a substantial stake than to affiliates in which they have a small share.[53]

Similarly, restriction of foreign ownership not only limits the flow of equity capital but may also reduce the joint venture's access to parent-supplied debt capital and short-term trade credit. And the parent firm may be more willing to retain profits at the subsidiary level when it has a large equity share in the subsidiary than when it has a small share.[54] To the extent that funds to replace foreign parent debt, trade credit, and equity are either not available to the joint venture from other sources or can be raised only at higher cost, restriction of foreign-parent share will result in a net loss to the host country.

The degree to which the host country will have to compromise on the question of ownership and control of joint ventures depends not only on the resources the country needs but also on the number of alternative sources from which these resources can be obtained. Clearly, when the needed technology is complex, rapidly changing, or possessed by only a few firms, the host country and local partners will have little bargaining power. In these circumstances, a deal can be struck only on ownership terms close to those dictated by the foreign firm. On the other hand, when the technology is standardized and widely available from many firms, the host country will be able to insist on a greater degree of local ownership and control.

The host country may also have some latitude in dealing with individual foreign firms. Companies differ from one another both in their desire for control over their subsidiaries and in their ability to achieve the degree of control they desire. The host country can benefit from a careful analysis of firms to determine which ones are most likely to agree to favorable terms.

Part 3

Management of Technology

11

R&D Abroad by U.S. Multinationals

ROBERT C. RONSTADT

Many multinational enterprises adapt their technology when it is transferred abroad; some generate new technology in their foreign operations. Both for adaptation of existing technology and for the generation of new knowledge, a few multinationals have established research and development units overseas.

U.S. multinationals make four distinctive kinds of R&D investments abroad, each serving a particular purpose. Over time, however, all four types of R&D investments tend to depart from their original charter and to follow a common pattern of evolution.

This chapter was adapted from the author's "International R&D: The Establishment and Evolution of Research and Development Abroad by Seven U.S. Multinationals," *Journal of International Business Studies,* Spring–Summer 1978, by permission.

This evolutionary pattern of R&D investment abroad has important implications for multinational managers, public policy makers, and theoreticians concerned with U.S. foreign direct investment.

This chapter summarizes an extensive investigation of the research and development investments made abroad by seven U.S.-based multinational organizations. Findings are presented about the factors affecting the initial investment in and the evolution of fifty-five R&D units established abroad by the Exxon Corporation (its energy businesses), the Exxon Chemical Corporation, International Business Machines (IBM), the Chemicals and Plastics Group of the Union Carbide Corporation, CPC International, Otis Elevator Company, and the Corning Glass Works.

R&D ABROAD: THE STUDY'S PURPOSE

In 1974, the seven U.S. multinationals in this study spent between 8 percent and 45 percent of their total R&D budgets abroad. These expenditures were allocated to forty-nine existing R&D units that had been created or acquired abroad to perform research and development activities (see table 11–1). Although six fewer R&D units existed in 1974 than had been originally established abroad, R&D expenditures abroad had been consistently increasing and indeed growing faster than domestic R&D expenditures.[1]

The purpose of the study reported in this chapter was to provide a better understanding of these increasingly important R&D investments by addressing four questions: What has been the form of these R&D investments? Why have they been made abroad? How have they evolved over time? And what is their significance?

THE FORM OF R&D INVESTMENTS ABROAD

Between 1931 and 1974, the seven U.S. multinationals made fifty-five R&D investments abroad, resulting in the establishment of fifty-five organizationally distinct R&D units.

Of the fifty-five R&D investments, forty-two were created or newly formed by the seven organizations, whereas thirteen were acquired by them when they acquired or merged with other companies. Table 11–2 presents a breakdown of these R&D investments for each multinational organization.

TABLE 11–1

R&D INVESTMENT ABROAD IN 1974, SUMMARY STATISTICS FOR SEVEN U.S. MULTI-NATIONALS

COMPANY	R&D ABROAD AS A PERCENT OF TOTAL R&D EXPENDITURES	NUMBER OF R&D UNITS	
		Established Abroad	Still Existing in 1974
Union Carbide Chemicals and Plastics Group	8%	13	11
Corning Glass	9	5	4
Exxon Chemicals	23	6	5
Exxon (energy businesses)	25	5	5
IBM	31	9	9
CPC International	38	8	6
Otis Elevator	45	9	9
Total		55	49

SOURCE: Company records.

TABLE 11–2

CREATED AND ACQUIRED R&D INVESTMENTS ABROAD

COMPANY	R&D INVESTMENTS		
	Created	Acquired	Total
Union Carbide Chemicals and Plastics Group	8	5	13
Corning Glass	3	2	5
Exxon Chemicals	5	1	6
Exxon (energy businesses)	5	0	5
IBM	9	0	9
CPC International	7	1	8
Otis Elevator	5	4	9
Total	42	13	55

SOURCE: Company records.

All thirteen acquired R&D units were "incidental" acquisitions, in the sense that they were simply part of an organization that was acquired primarily for assets other than R&D. In no case had a U.S. multinational acquired the organization in order to gain access to its R&D resources.

Extensive interviews with managers who were familiar with these fifty-five R&D investments, supplemented by analysis of data for each R&D unit, showed that they fell into four categories, according to their primary purpose when they were created or acquired:

1.   *Transfer Technology Units (TTUs).* R&D units established to help certain foreign subsidiaries transfer manufacturing technology from the U.S. parent while also providing related technical services for foreign customers.

2.   *Indigenous Technology Units (ITUs).* R&D units established to develop new and improved products expressly for foreign markets. These products were not the direct result of new technology supplied by the parent organizations.

3.   *Global Technology Units (GTUs).* R&D units established to develop new products and processes for simultaneous—or nearly simultaneous—application in major world markets of the multinational organizations. (In this particular sample, the United States is included.)

4.   *Corporate Technology Units (CTUs).* R&D units established to generate new technology of a long-term or exploratory nature expressly for the corporate parent.

Table 11–3 classifies the foreign R&D units into these four types. The data reveal that: 1) most R&D investments (56 percent) were TTUs created abroad; 2) very few ITUs were established abroad, and only two were created; 3) all acquired R&D investments were either TTUs or ITUs;[2] 4) all R&D investments in GTUs and CTUs were created or newly formed investments; 5) GTUs represented 9 percent of total R&D investments; and 6) CTUs were the least frequent, at 7 percent of total R&D investments.

All the GTUs in this study were created by one U.S. multinational (IBM). It seems likely that a larger sample would reveal that CTUs are more often established than are GTUs.

TABLE 11–3

R&D Investments Abroad Classified by Their Primary Purpose When They Were Created or Acquired

Units	All R&D Investments	TTUs	ITUs	GTUs	CTUs
Created units	42	31	2	5	4
Acquired units	13	6	7	0	0
Total	55	37	9	5	4

SOURCE: Company records and interviews.

RATIONALE FOR R&D INVESTMENTS ABROAD

Since the *acquired* R&D units in this sample had not been deliberately sought as such, we may focus on the motivation for creating R&D units abroad.

In almost every instance, the parent organization established an R&D unit for reasons directly related to the performance of the R&D function. Non-R&D goals—such as monitoring foreign R&D activities, taking advantage of "cheap" R&D labor, or using "trapped" or "blocked" funds—played no part in the vast majority of investment decisions and were of secondary importance in only a few instances.[3]

If non-R&D reasons were not significant, what factors were important for the creation of R&D abroad? Rationales and decision processes differed considerably among the four kinds of R&D investment.

Transfer Technology Units (TTUs)

According to interviewees, the decision process leading to the creation of Transfer Technology Units involved three steps:

1. The U.S. parents decided to make an investment in production operations.
2. Foreign managers in charge of production and marketing operations discovered that the product/process technology was unsettled.
3. Foreign managers foresaw an ongoing stream of technical service projects.

If these conditions did not exist, technical requests were handled by U.S.-based R&D units. For instance, a temporary R&D mission might be sent abroad when foreign managers expected only infrequent projects. When foreign operations were based primarily on U.S. exports, managers located abroad were reluctant to establish a permanent R&D unit, since the export faucet might be turned off unexpectedly.

The characteristics of the TTUs created abroad were consistent with the decision process described by the interviewees. All thirty-one TTUs were administratively linked to manufacturing affiliates, and twenty-six (84 percent) were located at manufacturing sites. However, the TTUs were located in only nine of the forty countries in which the seven parent companies had established production operations (see table 11–4). Most TTUs (61 percent) were located in Great Britain, West Germany, and France; these TTUs were the oldest R&D investments, with an average age (in 1974) of twenty-six years.

The small investment needed to create a TTU would not have prevented their establishment in other countries with manufacturing operations. The vast majority of TTUs (87 percent) had six or fewer

TABLE 11–4

LOCATION OF TRANSFER TECHNOLOGY UNITS BY COUNTRY

COUNTRY	NO. OF UNITS	PERCENTAGE[a]	CUMULATIVE PERCENTAGE
Great Britain	7	23%	23
Germany	6	19	42
France	6	19	61
Italy	2	7	68
Switzerland	1	3	71
Belgium	2	7	77
Netherlands	1	3	80
Canada	4	13	93
India	2	7	100
Total	31	100%	

SOURCE: Company records.
[a]All percentages rounded.

R&D professionals at the time of their creation. These were almost entirely host nationals already employed by the foreign subsidiaries, which used existing facilities to perform small, low-risk projects.

In the few instances where Transfer Technology Units were created outside the major foreign markets in Europe and Canada, the country involved had the potential for a large market with unique characteristics—e.g., Union Carbide's investments in agricultural chemicals in India. In addition, most TTUs in non-European nations were established after 1965, since these markets were developed later than their European and Canadian counterparts.[4]

Several managers noted also that manufacturing subsidiaries in Latin America, Africa, and the Far East—all of which had smaller markets than those of the three European nations and Canada—did not require permanent R&D units to perform technology transfer services. Product or process technologies had been standardized by the time manufacturing investments were made in these countries. When technical problems arose, they were handled by temporary R&D teams sent over from the United States or Europe.

Indigenous Technology Units (ITUs)

The decision process resulting in the creation of ITUs involved four conditions, according to interviewees:

1. Substantial foreign investment had been made in production operations.
2. The stream of new products and processes transferred from the U.S. parent had declined (or was perceived to be about to decline), and the future stream of new and improved products developed in the United States would be insufficient for the growth needs of the foreign business.
3. Foreign managers had identified new investment opportunities that were different from existing U.S. businesses.
4. Foreign managers were able to demonstrate to senior U.S. managers that there existed abroad the managerial and technical capability to implement new product development associated with a particular investment; consequently, foreign managers were able to secure permission for projects of

considerable size and scope that represented a major shift in the policy that all new product development would be performed in the United States.

The decision process just described resulted in the formation of two ITUs, one each by Otis Elevator and the Corning Glass Works. Both Otis and Corning managers stated that, until 1960 and 1970, respectively, new products developed expressly for the United States market were sufficient to sustain the growth of their companies' subsidiaries in Europe. Thereafter, both organizations found it more difficult to continue growing in Europe, as European market needs increasingly differed from the U.S. experience. As the European business expanded, European managers identified new investment opportunities that could not be exploited by using the technology developed by the parent companies. An Otis manager remarked:

> The president of the French subsidiary finally convinced corporate managers in the early sixties that Otis was missing out on a major market by not going into the small elevator business for low-rise buildings. The main reason Otis had not entered the market was that it was not in the business in the United States and, consequently, it did not have existing product/process technology to transfer abroad. As a result, the decision was made to let Europe develop its own products and production processes for the small elevator business in Europe.

A similar situation was described by a manager who had operating responsibility for Corning in Europe:

> For a long time, the European business grew rapidly, based on products developed by Corning in the United States. We were moving along like a car speeding at 100 miles per hour. No one was interested in or even had time to be concerned with new products designed specifically for Europe. But then managers saw growth slow down, and the car began moving along at 50 miles per hour and seemed to be ready to slow down even further. Several European managers became concerned and started suggesting that Corning/ Europe should begin developing its own products.

Only when managers decided that the flow of new U.S.-developed products would not be sufficient to sustain the continued growth of their businesses did they begin to pursue new investment opportunities that entailed new or improved product developments.[5]

Corning and Otis managers suggested that the investment opportunities they selected had to be sufficiently different from U.S. business interests to justify forming Indigenous Technology Units. Otherwise, the parent companies could have made a case for the domestic development of new products, since appropriate R&D resources, along with potentially larger markets, would have already existed in the United States.

In both cases, foreign managers noted that they had to have the approval of their U.S. parent companies to create Indigenous Technology Units, for a number of reasons. First, new product development abroad represented a distinct change in established policy. Both new and improved product development had traditionally been the responsibility of domestic R&D and engineering units. Second, when the decision involved a major product/market diversification (as it did for Otis), this step had to be reviewed carefully by corporate headquarters. Third, the investments in Indigenous Technology Units were substantial. Investment in R&D personnel was high, for example, because the commercial development of new products and processes tended to be much more complex than technical service activities. Building a pilot plant, tooling, manufacturing start-up, and the marketing of new products called for specialized skills and large-scale projects.

Also, the managers recommending the creation of Indigenous Technology Units had to have political clout within their organizations. In fact, the units analyzed in this study were initiated by European regional executives and the general managers of the largest European subsidiaries. Both ITUs were located in France, a country with well-established major markets and manufacturing operations that served both France and other nations. France was also the site of each organization's regional headquarters for Europe.

Global Technology Units (GTUs)

GTUs were created only if three conditions were met, according to interviewees:

1. R&D operations could be associated with substantial production and marketing resources already located abroad.

2. A total enterprise strategy required the production of a single product line for worldwide markets, as opposed to fundamentally different products for major world markets.

3. It was necessary to assign R&D responsibility for the development of particular products and processes to certain foreign subsidiaries, because the organization's innovative resources in the United States were fully utilized.

One U.S.-based multinational organization—International Business Machines—created five R&D units to develop new products for simultaneous, or nearly simultaneous, production in the U.S. and other foreign markets.

IBM managers noted that the five Global Technology Units were not established until the 1960s, when IBM decided to produce its 360 product line for the worldwide market, rather than a line of U.S. computers, a line of French computers, etc. Since IBM managers then decided, for competitive reasons, that the 360 line should be developed as quickly as possible, all nine new computer models were introduced at the same time. Consequently, the System/360 project, one of the most ambitious and innovative ventures in commercial history, required nine central processing models, new components technology, more than seventy new peripheral machines, and completely new software. The total cost was more than $5 billion, with $500 million spent on R&D alone. As T. A. Wise has written:

> It was roughly as though General Motors had decided to scrap its existing makes and models and offer in their place one new line of cars, covering the entire spectrum of demand, with a radically redesigned engine and an exotic fuel.[6]

The radical nature of the program spurred management to play an active role in the development of the System/360 and, subsequently, the System/370. During the 1962–1970 period, the company developed large and complex sets of new products for these two systems. This meant complex R&D assignments which, in turn, called for individuals with particular skills. Jobs became more numerous, more specialized, and less flexible.

The decision of IBM's senior management to assign responsibility for the initial development and production of some new products in the new line to certain foreign subsidiaries was made because the System/360 project exceeded the capacity of existing domestic development resources, not only in terms of R&D but also in the areas of manufacturing, engineering, and marketing. Furthermore, IBM did not consider it practical to expand domestic resources to handle all new product work. Growth ceilings placed on major manufacturing and marketing centers in the United States prohibited such large-scale expansion of existing operations. The managers who were interviewed noted that early in the company's history corporate leaders had imposed ceilings on the number of IBM personnel employed in any one setting. These limitations were based on the number of manufacturing, marketing, and R&D personnel relative to the total work force of a surrounding community.

According to senior IBM managers, these ceilings led to the creation and expansion of new centers of operations in the United States and abroad. IBM did not think it made economic sense to start new manufacturing centers in the United States solely for product development purposes, especially since manufacturing and marketing resources already existing abroad could be used for product development work once R&D units were created at these sites. Conversely, new product development abroad was highly improbable until the foreign manufacturing, engineering, and marketing functions existed, since U.S.-based multinationals as a group (and IBM in particular) did not make initial foreign investments to develop new products.[7]

As noted earlier, companies have found that effective product development work requires close personal interaction between R&D and other functional groups responsible for developing the new product.[8] The managers interviewed at IBM stated that the Global Technology Units were located at foreign centers of manufacturing and marketing because close administrative and geographic ties insured efficient communications. Nearly all IBM R&D laboratories were closely associated with marketing, engineering, and manufacturing units in the United States and abroad. Four of the five GTUs were located in or near a manufacturing center. One GTU in Sweden was located at a marketing center and was responsible for software development; consequently, the products developed at the Swedish unit could be immediately transferred abroad without a manufacturing function.

Corporate Technology Units (CTUs)

Interviewees stated that CTUs were created abroad when the following conditions occurred:

1. Corporate managers felt they must start or expand exploratory research of a long-term nature in order to protect the competitive position of their organization in the future.
2. Corporate managers decided to recruit top scientists located abroad.
3. The top scientists that corporate managers sought would not move to the United States.

Union Carbide, CPC International, and IBM created four R&D units abroad to generate new technology expressly for the corporate parent. Several interviewees at all three multinationals stated that CTUs were established during the mid-1950s and the early 1960s, when top corporate executives were worried about the long-run technological competitiveness of their organizations. This concern was particularly strong within Union Carbide and IBM, which made relatively large exploratory research commitments abroad, but it was also felt by CPC, which made only two small foreign R&D investments.

In IBM's case, it was Thomas Watson's concern about the company's frequent noncompetitiveness in technology that led the company to establish its Research Division in 1956. The new division's mission was to perform all long-term R&D activities—i.e., projects with sales impact five years or more in the future. This left short-term R&D operations to the operating divisions.

Union Carbide also made extensive investments in exploratory research in the United States at the corporate level while it created a similar unit abroad. Executives thought that R&D productivity was declining over time because of an insufficient amount of exploratory research in the operating divisions.

The decision to create CTUs abroad rather than in the United States was initiated when top managers shifted their focus from the U.S. scientific community to foreign scientists, because of scientific advances made by these scientists that were relevant to their businesses. Several managers mentioned that corporate managers in all three organizations knew of the work of foreign scientists before they invested in the Corporate Technology Units. For example, IBM's Swiss unit was formed in 1956, shortly after several European scientists had performed work on memory cores that had a revolutionary impact on the computer industry.

The final step in the decision to create a CTU abroad occurred when corporate managers found that they could not always persuade foreign scientists to come to the United States to perform exploratory research. These projects typically required considerable time, and a continuity both within and between projects was also important. Consequently, a foreign scientist moving to the United States would have to expect to remain several years.

All four CTUs were formed between 1956 and 1960, when corporate management in the companies in the study were deeply concerned about the need to generate new technology. Initially, each CTU's R&D expenditures were oriented toward exploratory research of a long-term nature. A number of R&D managers noted that a long-term charter was given to all four units, with the goal of generating revolutionary concepts. One manager remarked: "Top corporate executives have invested in us to come up with some surprises that are relevant to the company. They don't expect us to come up with something that they can sell tomorrow."

In all four cases, the CTUs were geographically and administratively separate from other foreign investment operations, because of the long-term, exploratory nature of the R&D activity. Managers preferred to preserve the independence of the CTUs in order to prevent operating units from diverting attention to shorter-run problems. In each case, the R&D directors of the CTU reported to superiors at the corporate level.

Managers stated that when CTUs were large (more than twenty R&D professionals), they were located in nations like Switzerland or Belgium, countries close to other European nations where top scientists could be recruited. The two CTUs that were very small (only a few R&D professionals) were purposely located in the home countries of the leading scientists with skills relevant to the companies' needs (Italy and Japan).

Whether the units were large or small, the choice of a location was based on the need to attract highly skilled professionals from a given geographic area. The nations in which the CTUs were located—Italy, Belgium, Switzerland, and Japan—were not large markets for the parent organizations.[9]

EVOLUTION OVER TIME

The evolution of R&D investments abroad has shown two basic patterns:

1. A tendency to change purpose and to continue operations at the same location, and a related tendency to increase the number of R&D professionals considerably when a change of purpose occurs; and

2. A tendency toward slow growth or divestment if a change of purpose does not occur; divestment results in consolidation with other R&D units if possible, or in complete disbandment.

The forty-two R&D investments created abroad tended to change their purpose to developing new and improved products and processes expressly for particular nations or regional foreign markets, regardless of their original purpose. For example:

• Most TTUs have evolved or are evolving into developing new or improved products for particular foreign markets.

• The two ITUs recently created abroad continue to perform similar work for particular foreign markets.

• A number of Global Technology Units are also performing increasing amounts of indigenous technology work.

• Two CTUs were disbanded when they produced no significant results and could not change to some other R&D function. Of the two remaining CTUs, one was evolving completely into indigenous technology work for the Japanese market; the other was providing more research support for European development efforts.

As shown in table 11–5, there were eleven ITUs in 1974, nine more than had been originally created. In each of these nine cases, the

TABLE 11-5

R&D Investments Created Abroad, Classified by Their Primary Purpose

Units	All R&D Investments	R&D Investments in:			
		TTUs	ITUs	GTUs	CTUs
When created	42	31	2	5	4
In 1974	38	17[a]	11	8[b]	2[c]

SOURCE: Company records and interviews.
NOTE: The thirteen R&D units acquired abroad by the sample firms are excluded from this table.
[a]Several other units were increasingly involved in indigenous technology work and were expected to evolve into ITUs in the near future.
[b]Includes three units that had evolved first into indigenous technology work and then into global technology work.
[c]Includes one unit expected to evolve into indigenous technology work.

R&D unit had been created as a TTU but had evolved into an ITU for two reasons: R&D directors of these units felt they had to provide more challenging work to keep their best people and to attract high quality R&D professionals; and marketing and general managers abroad felt an increased need for new products to maintain growth in sales and profits.

Managers stated that TTUs needed additional personnel because existing customers with changing product needs required technical service work. Exxon R&D managers mentioned, for example, that Mercedes-Benz requested new types of automotive oils and lubricants whenever the company changed its engine design. In addition, new customers with product needs required technical service assistance.

Several managers also mentioned that, as Technology Transfer Units matured, their staffs grew less rapidly. A unit could handle a moderate number of new projects (three to five) by adding only one new R&D professional. Since the R&D staff became more efficient as it accumulated learning within a relatively narrow technological field, it could handle more projects and solve problems faster. A number of managers indicated that, in most cases, new products were introduced frequently enough to forestall reductions in the number of R&D professionals or the termination of a unit. But the introduction of new projects was not frequent enough to justify large increases in R&D personnel. In fact, three of the TTUs in this study were terminated and their activities were consolidated elsewhere. However, all three units had continued to grow until they were dissolved.

In 1974, seventeen TTUs continued to perform transfer technology services, while fourteen TTUs had evolved into Indigenous Technology Units. In this latter group, eleven units were still performing indigenous technology work in 1974; three units (all at IBM) had changed their primary R&D purpose for a second time in the early 1960s and had become Global Technology Units. (Later, as has been mentioned, IBM *created* five new GTUs.)

The data on average size, average age, and average yearly growth for the thirty-one TTUs fall into different patterns according to the units' changing purposes. For example, the eleven TTUs that changed their primary R&D activity to indigenous technology work were larger, older, and growing more rapidly than those TTUs that did not change their function. Similarly, the three R&D units that evolved into Global Technology Units were considerably larger and older than the ITUs and were also adding more R&D professionals per year (see table 11–6).

Managers observed that when TTUs became ITUs they increased their professional staffs, because indigenous technology projects were larger than technical service projects. A few managers noted, however, that after their units had been involved in indigenous technology for some time, the increase in the number of R&D professionals

TABLE 11–6

Selected Characteristics of R&D Units Originally Created as Transfer Technology Units (Classified by Their Primary Purpose in 1974)

	Still TTUs	Now ITUs	Now GTUs
Number of R&D units (total = 31)	17	11	3
Average size in 1974 (number of R&D professionals)	19	46	500
Average annual increase in R&D professionals[a]	1.17	1.87	12.1
Average age (years)	13	23	>35

SOURCE: Company records and interviews.
[a]From date of creation until 1974.

declined. A few interviewees stated that there was a fast build-up of R&D resources when operations began at a regional level, in order to develop new or improved products and processes for a larger market. But once these resources had been expanded, the increase in staff slowed down.

The data reinforced the notion that change of purpose influenced the *level* of growth in the number of R&D professionals. In all cases, the size of the professional staff increased as a function of R&D unit age, but the absolute level of employment depended on the type of R&D unit involved. The pattern is that of a step function, as shown schematically in figure 11–1.

This notion of a step function associated with a change of R&D

Fig. 11–1.   Schematic depiction of the purpose of R&D units created abroad, classified by age and the number of R&D professionals.

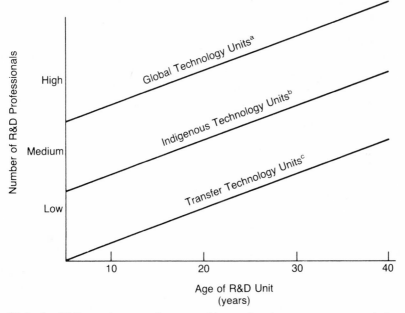

[a]Defined as R&D operations to produce new and improved products or processes expressly for major world markets.
[b]Defined as R&D operations to produce new and improved products or processes expressly for a particular national or regional market.
[c]Defined as R&D operations of a technical-service nature but based on technology supplied by the multinational parent.

purpose is also supported by the observation that the increase in the number of R&D professionals was larger for "younger" units; i.e., for those units that were created as Transfer Technology Units or that evolved into Indigenous Technology Units *after* 1960 (see table 11–7). These data also support the inference that the increase in the absolute number of R&D professionals tended to decline over time, until an R&D unit changed its purpose.

Earlier analysis of the creation of R&D units abroad showed that "older" R&D units were located in large-market nations.[10] Thus, the present study would suggest that the increase in the number of R&D professionals at "older" Transfer Technology Units and at Indigenous Technology Units located in Great Britain, France, and West Germany should be smaller than increases experienced by the same kinds of "younger" R&D units located elsewhere. In fact, the available data on geographic location show that Transfer Technology Units located in these three large European markets *have* made smaller staff increases, on an average annual basis (approximately the same average increases as shown in table 11–7), than similar R&D units located in other countries.

As noted earlier, three R&D investments that were originally created by IBM as Transfer Technology Units evolved into Indigenous Technology Units and later into Global Technology Units. The three R&D units were created as TTUs in the early 1930s and were located in Great Britain, France, and West Germany. Managers stated that the

TABLE 11–7

AVERAGE YEARLY INCREASE OF R&D PROFESSIONALS FROM DATE OF TRANSFER TECHNOLOGY UNIT'S CREATION TO 1974

	CREATED BEFORE 1960		CREATED AFTER 1960	
	No. of Units	Average Yearly Increase	No. of Units	Average Yearly Increase
Remained transfer technology unit	8	.98	9	1.4
Evolved into indigenous technology units[a]	5	1.7	6	2.1

SOURCE: Company records and interviews.
[a]Does not include three IBM units.

number of R&D professionals in these three units did not increase significantly until they started work on new product development for global application (projects associated with the System/360). The increase in the number of R&D professionals slowed down when the units reached capacity in facilities and in employment.

Subsequently (between 1967 and 1970), IBM created five new Global Technology Units: three in Europe, one in Canada, and one in Japan. The average staff size for these five younger GTUs was 143 R&D professionals in 1974, and they were adding an average of nineteen professionals a year. Unlike the three older GTUs in France, West Germany, and Great Britain, the five new units were created expressly to perform specific global R&D missions within specialized technological areas. The younger GTUs had to build their resources as quickly as possible within their specialized areas, in order to compete with older Global Technology Units (as well as domestic units) that had a broader range of technological skills.

This explanation is supported by data showing that the increase in the number of R&D professionals at the five younger R&D units created as GTUs was larger (24.7 R&D professionals per year for the two youngest units) than for the three older units (15.8 R&D professionals per year). All five GTUs were created with approximately the same number of R&D professionals, but when goals remained constant, staff growth apparently slowed with age.

There were no available company data concerning the increase in the size of the three older R&D units after they became Global Technology Units. However, conservative estimates suggest that increases in R&D professionals since the early 1960s average at least twenty per year, possibly as many as thirty per year.

Managers increased the number of R&D professionals at all eight GTUs because of their need to avoid project delays for a set of interrelated R&D projects that were closely tied to engineering, production, and marketing projects for the System/360 and, later, the System/370 product lines. Corporate headquarters coordinated project missions for domestic and foreign R&D units. They assigned projects on the basis of an R&D unit's relevant technological skills and the availability of resources.

However, IBM managers noted that the R&D activities of the three older Global Technology Units had begun to include more indigenous technology work when IBM altered its international marketing strategy. The company's new strategy called for greater specialization of hardware and software to meet specific user-industry needs. Because there were differences between customers in the United States and Europe—e.g., U.S. versus British banking—the company had to design product systems for particular national or regional industries. As the three older GTUs began changing their purpose, between 1971 and 1974, they assigned at least 300 R&D professionals to indigenous technology work.

Some IBM managers believed that, as R&D purpose evolved at the three older Global Technology Units, they would also become increasingly oriented to the development of new products expressly for a particular foreign market. Table 11–8 shows an estimate of the 1974 composition of R&D professional staff, by R&D purpose.

Of the four Corporate Technology Units, only IBM's Zurich unit and CPC's Japanese unit were still operating in 1974. According to interviewees, Union Carbide's Belgian unit was divested after fourteen years, and CPC's Italian unit disbanded after ten years, because they did not produce any significant transfers of technology to the corporate parent. Also, CPC's R&D unit in Japan did not produce any major flow of technology to the U.S. corporate parent; it survived, however, because corporate managers and general managers of the Far East affiliate successfully changed the R&D unit's purpose. Although poorly located in relation to the affiliate's manufacturing

TABLE 11–8

NUMBER OF R&D PROFESSIONALS IN IBM'S EIGHT GLOBAL TECHNOLOGY UNITS, 1974

	TRANSFER TECHNOLOGY WORK	INDIGENOUS TECHNOLOGY WORK	GLOBAL TECHNOLOGY WORK
Three units that evolved into global technology units	150	300	1,050
Five R&D units that were created as global technology units	65	25	625
Total	215	325	1,675

SOURCE: Interviews and author's estimates.

operations, the R&D unit was able to perform indigenous technology work for the Japanese subsidiary.

The two divested CTUs did not change their R&D purpose. Union Carbide management had worked hard for change of its Belgian unit but was unable to effect it. The interests of scientists at CPC's unit and its isolation from other CPC operations made any change of purpose infeasible.

Managers indicated that the poor performance of the three Corporate Technology Units was related to an insufficient allocation of R&D resources for exploratory research within a focused technological area. All three CTUs had small amounts of R&D resources allocated to single technological fields. The two CPC units had only four and six R&D professionals respectively. Union Carbide's Belgian unit employed 180 people and thirty-seven R&D professionals, but interviewees noted that these resources were divided among a number of unrelated or loosely related projects intended to develop different kinds of new businesses for Union Carbide that were unrelated to its existing businesses. The director of the Belgian unit confirmed that, throughout its existence, his unit performed only exploratory research to develop high-risk, unrelated businesses. Indeed, it may be that the Belgian unit was, in a sense, a collection of several small R&D units (unrelated project teams) within a common facility rather than an integrated operation.

By way of contrast, managers familiar with IBM's Zurich unit felt that it had survived as a CTU and increased its R&D personnel because of its larger allocation of R&D resources for exploratory research within specified technological fields that supported IBM's existing businesses. However, IBM managers observed that, even with this focusing of resources, the Zurich unit would not have been successful had it not shared projects with the larger U.S. exploratory research unit. The consensus was that, by sharing work with the Yorktown (New York) research unit, the Zurich unit was able to reach the critical size needed for success. It was the only CTU able to identify specific and important flows of technology to the U.S. parent.

Despite its success, IBM's CTU changed its purposes somewhat, though in 1974 its primary work was still in the area of corporate technology. The unit increased its number of R&D professionals at an average annual rate of 2.5; IBM managers noted, however, that part of this increase was associated with exploratory research, in support of indigenous technology work undertaken by the French R&D unit during the early 1970s.

THE SIGNIFICANCE OF THESE FINDINGS ABOUT R&D INVESTMENTS ABROAD

Significance depends in part on your vantage point—whether you are a manager of a multinational enterprise, a U.S. public policy maker, a foreign public policy maker, or a researcher interested in theory. Yet policy should be grounded in theory and generally is, as Keynes noted some time ago, whether or not the policy maker realizes it. So let us begin with theory—and move gradually into the realm of policy.

For theoreticians, the study's critical findings are that specific kinds of R&D investment exist and that the vast majority of foreign R&D investments seem to follow a particular evolutionary pattern. This R&D pattern or process begins with small investments made first in technical service labs in order to help transfer U.S. technology. Later, these R&D investments usually expand and evolve into organizational units seeking to develop new and improved products and processes expressly for foreign markets. Again, given the proper preconditions, these R&D investments may expand further and evolve into organizational units seeking to develop new and improved products and processes for simultaneous manufacture in several major world markets.[11]

From a theoretical perspective, the creation of R&D investments to perform global technology work has important ramifications for scholars who are interested in theories of foreign direct investment. It has particular relevance for the partial model known as the Product Life Cycle (PLC) theory of U.S. trade and foreign direct investment.[12]

The study's findings about R&D investments made abroad to help transfer U.S. technology (TTUs) are consistent with and support the PLC model. Also, the findings about R&D investments to develop new and improved products and processes abroad expressly for specific foreign markets (ITUs) present no theoretical conflicts with the PLC model. However, the creation of R&D investments to help develop new products and processes for simultaneous manufacture in and commercialization for major world markets (GTUs) moves beyond the explanatory and predictive powers of the PLC model. A new scenario—or at least an expansion of the existing one—is required, since the model does not account for the development abroad, for initial commercialization simultaneously in the United States and major foreign markets, of new high-income products that are also labor saving (e.g., computers).

Are Global Technology Units an IBM anomaly, or will other, larger U.S. multinationals perform this kind of R&D work abroad in the future? It seems likely that we will see more GTUs—and if so, the product cycle model will be less useful in explaining the international investment activity of major U.S. multinationals.

For multinational managers, these findings have three important administrative implications. First, different kinds of R&D investments exist and must be analyzed according to their individual purposes. Corporate managers should consider the evolutionary possibilities of foreign R&D operations early in the process, given the tendency of these operations to move in specific (albeit unplanned) directions in the absence of corporate involvement and to add R&D professionals rapidly when the unit changes its purpose.

Second, corporate managers should carefully evalute R&D units created expressly to generate technology for the corporate parent, given the adverse experiences of this type of unit among those in the present study.

Third, the data on evolution suggest that, regardless of their original purpose, R&D investments should be flexible in terms of purpose. The conditions that led to their creation will change, and R&D investments must be capable of adapting to new requirements. If they are not, the R&D investments that will fail are those that, for example, are isolated from other operations or are located in "neutral" nations where no operating subsidiary exists, or those that cannot be consolidated with other R&D activities abroad.

For U.S. policy makers, the findings indicate there are good economic reasons for U.S. multinationals to perform R&D abroad. The findings also suggest that R&D investments, especially in TTUs, ITUs, and GTUs, have made U.S. multinationals more competitive in foreign and U.S. markets than they would have been if they had performed R&D only in the United States.

Flexibility will also be important for foreign policy makers where

R&D investments exist or are likely to be created. For instance, a foreign policy maker may desire only ITUs that will develop new and improved products expressly for the national or regional market. Yet ITUs may take time to evolve and, once they exist, may need to change their purpose again to adjust to a new corporate strategy.

The data also suggest that foreign policy makers will be unable to attract investments in R&D unless good economic reasons exist for the investment. Here, the foreign policy maker—no less than the multinational manager—must understand the process of R&D investment in order to know when these economic conditions exist. Most R&D investments that are created abroad are associated with certain kinds of manufacturing investments, where product and process technology are unsettled. And these R&D investments will be small, low-cost commitments for transfer technology work.

However, foreign policy makers have an interest in helping to speed the evolutionary process of TTUs toward indigenous technology work, since such a shift creates employment and provides new products and processes developed expressly for the local market. In this regard, a positive investment climate and, possibly, incentives for particular kinds of R&D work may be helpful to R&D directors of TTUs who are trying to sell projects to their supervisors, who, in turn, will have to persuade general managers located abroad to undertake R&D projects of larger scale.

12

Influence in the Multinational Enterprise: The Case of Manufacturing

HENRI DE BODINAT

When a firm transfers technology abroad, it usually exercises some control over the receiving unit. The extent of that control helps determine the impact of the transfer on the host country, according to Rafii's study (Chapter 10). But to date little has been known about exactly how the multinational imposes its will on the foreign entity. Its influence is, according to this study, the result of a complex process.

Influence is an elusive but frequently invoked concept. The fashion in Paris is said to influence the fashion in New York. The United States influences Saudi Arabia, and vice versa. Lawrence

This chapter is based on the author's "Influence in the Multinational Corporation: The Case of Manufacturing," D.B.A. dissertation, Harvard Business School, 1975.

Sterne has influenced Norman Mailer. A very broad definition of *influence* might be the power to exert a force at a distance in time or space. In a social system or an organization, the definition should be narrower; the definition of influence used in this study is the power to stabilize or modify the behavior of an individual or an organizational unit.

Influence is an important phenomenon. As Leavitt says, "It is not only a problem which pervades business and industry . . . it is also a central problem in the family, in education, in psychiatry, in international relations, in politics, and in every other phase of human interaction."[1] Seldom, however, has the phenomenon of influence been directly tackled by researchers. The literature on systems theory, sociology, organizational behavior, management control, international business, and business policy includes only sparse clues as to the nature of influence and even less explanation of the phenomenon.[2] The research reported here was undertaken to help fill this gap.

This study examined influence in the setting of the multinational firm. Multinational firms, a fascinating subject in their own right because of their growing economic and social importance, are also an excellent example of complex systems. The separation of the parent's headquarters from the foreign subsidiaries makes it easier to observe the process of influence.

The manufacturing function was chosen as the study focus. This function illustrates the problems and opportunities of multinational firms but has not been extensively researched.[3] In line with the concepts used in manufacturing management courses at the Harvard Business School, the subject was split into three general types of decisions, each with several subcategories:[4]

Major Decisions	Standards and Procedures	Day-to-Day Management
Adaptation of process	Quality	Quality
Adaptation of product	Cost	Inventory
Capacity expansion		Production scheduling
Vertical integration		

A two-stage interviewing procedure was followed. First, in-depth interviews—requiring five full days in total in each firm—were conducted with executives in three firms (Corning Glass Works, Rhône Poulenc, and Unilever), including several locations for two of the firms. These in-depth interviews revealed considerable consistency of opinion among different executives of the same firm who were asked to describe, decision by decision, the respective inputs of the parent and of the subsidiary. This consistency made it possible to restrict the second-stage interviews to the parent or the subsidiary level only. These standardized interviews required about half a day at each of thirty firms. Most of these interviews took place in France, because of financial and time constraints and because it seemed acceptable on methodological grounds.

Of the thirty-three firms in the sample, nineteen were headquartered in France, thirteen in the United States, and one in United Kingdom/Netherlands.[5] A wide variety of firms—small as well as large—were included in order to see whether size and complexity have a strong bearing on forms and degrees of influence. In this research, a multinational corporation is defined as a firm of any size with at least one manufacturing subsidiary abroad. In fact, most firms studied had many manufacturing subsidiaries abroad.

Three general principles governed the research orientation. First, explore rather than describe. Second, explain rather than prescribe. And finally, build hypotheses rather than test them. The results of the study illuminate both the *nature* of influence (What is influence?) and the *explanations* of the phenomenon (Why is there influence and why does its intensity vary?).

THE NATURE OF INFLUENCE

The network of influence in a multinational enterprise is highly complex, even if one focuses only on parent-subsidiary relationships. This complexity arises first from the sheer multiplicity of goals, methods, channels, and directions of influence, but it also reflects other characteristics of the phenomenon, such as ambiguity and heterogeneity.

Multiplicity of Goals

Field research indicates that the parent's main goal in exerting influence is to increase the contribution of a subunit to the well-being of the whole organization. This general goal might itself be broken down into three subgoals. A parent influences a subsidiary in order to transfer knowledge to the subsidiary, reduce slack in the subsidiary's operations, and coordinate the subsidiary with the other subsidiaries of the system.

The experience curve concept suggests the motive for the transfer of knowledge: If knowledge is transferred between two manufacturing subsidiaries, each will be able to use the experience, due to the total of their accumulated output, and thus to lower its own costs.[6] In this case, influence based on knowledge can proceed not only from a more sophisticated to a less sophisticated unit, but even between units (the parent and the foreign subsidiaries) of equal sophistication, and from a less sophisticated to a more sophisticated unit. Some firms have established formal structures to transfer knowledge. For instance, the Unilever toiletry and health-care product divisions store in a central computer memory all the latest information on processes used or products made by any foreign plant, and every subsidiary has free access to this data bank.[7]

The second goal of influence is to reduce slack within a subunit. Slack might be defined as the difference between the actual efficiency of a subunit and the efficiency it could achieve. At Unilever, standard toiletry product costs are proposed each year by the French subsidiary, Gibbs, using a format determined by the parent. In the words of one of the Gibbs manufacturing managers, "The product division in London always says, in reaction to our preparation of standard costs: 'They are too high.'" Similarly, at IBM, a series of ratios are computed for every plant, and the foreign subsidiaries are expected to come as close as possible to the best ratio (i.e., the ratio of the subsidiary that performs best in this respect).

The third purpose of influence is to improve the coordination between subsidiaries. For instance, the Corning parent corporation decided that its French subsidiary, Sovirel, should not invest in a new glass furnace to serve the Japanese market, because the Japanese subsidiary was judged better equipped to play this role. Similarly, IBM has imposed extremely strict rules of inventory management on the plants that sell components to other plants.

To reach these three goals, a firm can use a wide portfolio of methods and styles of influence.

Methods of Influence

The field research indicated three principal methods of influence: direct influence, indirect influence through control systems, and indirect influence through acculturation.

Direct influence is exerted when the parent company directly intervenes in a major decision that concerns a foreign subsidiary, such as the decision to expand capacity; or when the parent directly designs a standard or procedure that the subsidiary will have to apply, such as a production scheduling technique; or when the parent gets involved in the day-to-day management of the subsidiary, for instance by determining the inventory level.

Along the continuum running between total centralization at the parent's level and total freedom of the subsidiary, the in-depth interviews suggested five main patterns of direct influence:

1. *Absolute centralization.* The parent decides alone. The subsidiary is not even entitled to initiate a proposal or to give its advice.
2. *Participative centralization.* The parent decides, but the subsidiary might give its advice or initiate a proposal.
3. *Cooperation.* Both the parent and the subsidiary have to agree: bargaining or cooperation.
4. *Supervised freedom.* The subsidiary decides, but the parent might give its advice or make a suggestion.
5. *Total freedom.* The subsidiary decides or manages alone.

This typology was then used as the framework for collecting data on all the firms studied.

Absolute centralization is fairly rare. For example, the choice of the production process and the extent of its adaptation to the local conditions was made on this basis in only four firms (13 percent of the thirty-one firms for which such information was available). The pattern was one of participative centralization in seven firms (23 percent), of cooperation in three (10 percent), of supervised freedom in thirteen (42 percent), and of total freedom in two (6 percent).

The dominant pattern of influence varies strikingly in the sample from decision to decision or from area of management to area of management. For the major decision of vertical integration of the foreign subsidiary, the dominant pattern is clearly centralization, in either its total or its participative form; twenty-three firms out of thirty-two (72 percent) followed this pattern. For product adaptation, however, freedom (either supervised or total) characterized the decision process for sixteen firms out of thirty-two. In day-to-day management, production scheduling is left to the subsidiary's initiative (supervised or total freedom) in twenty-two firms (67 percent).

On the whole, the subsidiary has more freedom in establishing standards and procedures than in its main manufacturing decisions, and even more freedom in its day-to-day management, as shown in table 12–1. The total freedom category applies in only 10 percent of main manufacturing decisions, 20 percent of standards and procedures choices, and 42 percent of day-to-day management.

Indirect influence through control systems undoubtedly exists, though its impact is difficult to evaluate precisely. All firms interviewed, for instance, asked their subsidiaries to set up a budget. But the intensity of indirect influence differed sharply from firm to firm. The following four degrees of intensity were observed.

TABLE 12–1

LEVEL OF PARENT'S DIRECT INFLUENCE ON MANUFACTURING DECISIONS OF FOREIGN SUBSIDIARIES, 33 MULTINATIONAL ENTERPRISES (PERCENTAGE OF TOTAL OBSERVATIONS IN EACH CATEGORY OF DECISION)

	TYPE OF DECISION		
	Major	Standards and Procedures	Day-to-Day
Centralization (absolute or participative)	47%	29%	19%
Cooperation	17	27	22
Freedom (supervised or total)	35	44	58
Total	100%	100%	100%
Number of observations	127	66	99

SOURCE: Author's interviews.
NOTE: Not all columns total 100 percent because of rounding.

1. The subsidiary's budget was considered as a mere forecast that the parent never discussed in four firms (13 percent of total).

2. The budget was a forecast discussed only briefly by the parent and not every time in ten firms (31 percent).

3. The budget was seen as a commitment but was automatically accepted by the parent in one firm (3 percent).

4. The budget was a commitment of the subsidiary that the parent negotiated with or imposed on the subsidiary in seventeen firms (53 percent).

Thus, the indirect influence exercised through the budget could be considered low in about half of the sample firms (the first three categories) and high in about half.

In general, indirect influence through central systems can be broken down into two steps: a "quasi-bargaining" step, and a "cybernetic" step. In the first step, the parent and the subsidiary negotiate the goals, or the budget. In the second, the parent monitors the meeting of these goals. In a firm such as Unilever, the standard costs of edible fats for the French plant are negotiated at the end of each year with the product division in Rotterdam. Then the actual costs are compared with the standards during the following year, and a final review takes place at the beginning of the next year.

Bargaining over goals concerns the level of performance that the subsidiary is supposed to meet for a future period. Two offsetting tendencies determine how the parent and the subsidiary reach an agreement. Because it becomes clear that the performance will be more and more difficult to realize, the subsidiary's resistance tends to grow as the budgeted performance increases (except for some pathological cases where the subsidiary does not care, perhaps because the system is not efficient). The parent's pressure, on the contrary, tends to decrease as the projected subsidiary performance becomes more and more satisfactory. A bargain is struck at the performance level at which the subsidiary's increasing resistance overcomes the parent's declining pressure. This pattern was observed at both Unilever and Corning Glass Works.

The feedback phase is more straightforward. It consists of the parent's checking whether the objectives are met and reacting if they are not. Curiously enough, most companies seem to put more emphasis on this second phase (feedback) than on the first (negotiating the budget). The first phase is diluted and foggy, with a bargaining process that is not clearly perceived, but the second phase is clear-cut. As a result, some erratic reactions can be observed in the first phase, especially if the parent's bargaining position is weak. At Corning, for instance, the budgets of all the subsidiaries' product divisions were accepted one year, and they were all refused the following year. This unpredictability in the bargaining phase makes it more difficult to carry out the feedback phase effectively. Under such conditions, the budget process can become useless, except for forecasting or the detection of major problems.

Indirect influence through acculturation is not yet clearly understood and consciously managed, but it seems increasingly important in the big multinational firms; some two-thirds of the firms in this study used it. The increasing importance of this form of influence is reflected, for example, in the rise of the executive manpower management function (as at Corning and IBM). IBM has developed a sophisticated executive manpower management system that allows a systematic rotation and in-house training of executives. In this way the corporate values and methods become imbedded in the minds of these executives, who can then reach appropriate corporate-wide decisions efficiently, without any influence exerted by the parent. This acculturation also facilitates communication between the units of the multinational firm: executives everywhere have a common code and language, and they know with whom they should communicate about what. Communication is thus smoothed and shortened.

One might, however, distinguish two types of acculturation at the foreign subsidiary's level. *Exported* acculturation (employed by 42 percent of the firms) consists of exporting former parent-company executives to the subsidiaries. This process is simple, represents a guarantee of acceptable performance by the subsidiary, and facilitates communication and influence between the parent and the subsidiary. But some problems of communication might then be nested *within* the subsidiary between the expatriates and the local nationals who are culturally (nationally and organizationally) different. (This was the case at Corning.) In *managed* acculturation (employed by 21 percent of the firms), the subsidiary's managers are local nationals with experience at the parent company. The parent's managers are a mix of parent-country executives with some international experience and foreign-country executives who have come to work at the parent's headquarters (area, product division, international division, or corporate levels).

Styles of Influence

Just as methods of influence vary, so do influence styles. Broadly, it is possible to distinguish two types of influence: authority and cooperation. Authority means that influence is mainly top-down, hierarchical, and unilateral. Cooperation means that influence is bottom-up as well as top-down, and the decision process stresses consensus. The Japanese *ringi* system can be viewed as a typical case of cooperation.[8] Leavitt has also mentioned coercion and manipulation;[9] but coercion was not observed in the firms studied here (probably because it was unnecessary), and manipulation is extremely difficult to observe and can be considered an interindividual idiosyncrasy rather than an organizational pattern.

We can apply this typology of styles to any face-to-face contact between two executives or two groups (as in the case of direct or indirect influence through control systems). But the typology cannot readily be applied when the process of influence is diffuse and lengthy, as in acculturation.

Both styles exist in the multinational companies that were interviewed in this study. For direct influence, pure cooperation, in which the parent and the subsidiary must agree on the decision, is the dominant pattern in six firms out of thirty-one ($\cong$ 19 percent). For indirect influence, however, cooperation is the dominant pattern in most firms, at least during the first step (bargaining, goal setting). Only two firms imposed goals on the subsidiary; all the others negotiated with the subsidiary or even accepted without discussion the subsidiary's proposal. In the second step (monitoring and identification of deviations), it was difficult to distinguish clearly a style of relationship, since a deviation was the signal for a set of progressive interactions with the parent that often began in cooperation and finished in authority.

Multiplicity of Channels

Influence proceeds through many channels. In addition to the formal vertical channels, which can be used in one direction only (top-down) or in both directions, there were informal influence channels and lateral channels.

The most striking example of informal influence channels might be found at Corning Glass Works. The product-division managers in the French subsidiary were formally connected to two chains of influence: one based on geography (subsidiary manager and area manager) and the other on product line (business manager). But for three of the four product-division managers, the most important contact in the parent company was none of these formal hierarchy representatives but the manufacturing managers at Corning domestic operations. They were in frequent contact with these managers, who in practice served as advisors in technical matters, gatekeepers to Corning's powerful informal networks, and manufacturing coordinators of the group decisions worldwide.

Several examples of lateral influence were also found. Unilever's subsidiaries in Europe were free to get in touch with each other directly on technical matters. And the French plant managers would call their counterparts in Germany, Holland, or Switzerland several times a year to discuss with them a new product or process, or to get their help in solving manufacturing problems. Similarly at Corning, the glass molding specialists of Sovirel (the French subsidiary) and Jobbling (the U.K. subsidiary) met yearly to compare their processes and management.

Characterization of Influence

Influence is thus a complex phenomenon, which is used for different purposes, can take different forms, and can flow through different channels. Influence is both heterogeneous and ambiguous.

Influence varies within functions (such as manufacturing) and within classes of decision. Direct influence, for example, is often used in the major manufacturing decisions, but the same firm might make decisions related to vertical integration in the framework of total centralization and decisions related to capacity expansion in the framework of supervised freedom.

The phenomenon of influence is ambiguous, making it difficult, for example, to determine just how belonging to a multinational firm has affected the product of the subsidiary. It is relatively easy to analyze the direct influence of the parent on a decision, although some of the words used in our typology, such as *advice* or *initiation*, are ambiguous. But when the influence is indirect, it becomes difficult to sort out the effects of the different sources of vertical influence (e.g., the area headquarters versus the product division). It is also difficult to assess the impact of informal or lateral influence. The potent but diffuse and foggy influence of acculturation gives a final touch of ambiguity to the picture.

The corollaries of these findings are three. First, a simple, cybernetic view of influence is inadequate, because it ignores reciprocity: influence can proceed from the bottom up. Moreover, influence can be lateral, flowing from subsidiary to subsidiary.

Second, the polar centralization/decentralization alternative is an unrealistic representation of a world in which other styles, such as cooperation, exist. This centralization/decentralization view fails to take into account the multiplicity of the methods of influence and of the classes of decisions in which the parent's influence might be exerted, as well as the inherent substitutability and complementarity of these methods and classes.

Third, the multiplicity of the directions of influence should remind one that the subsidiary is not a pawn on a chessboard but may influence the parent. Sometimes this influence may extend beyond decisions concerning the subsidiary itself to choices affecting the fate of the whole system. The brain is not confined to the parent level; intelligence and decision-making power may be scattered throughout the subsidiaries.

WHY MULTINATIONALS DIFFER IN TERMS OF INFLUENCE

Multinational firms differ considerably in the degree of influence they exercise over their subsidiaries abroad. The question of the parent's influence is a key point in the heated controversy about multinational firms. At the firm level, this is the problem that raises the most questions for the top executives of the multinational enterprise. Thus, it seemed worthwhile to explain some of the reasons for the observed differences between firms.

Method of Comparison

Through interviews a pattern of relationship between the parent and the subsidiary was established for each company, for every major manufacturing decision, for the important standards and procedures, and for the main day-to-day decisions. The basic method was then to compare two firms for each of the three methods of influence (direct, indirect through control systems, and indirect through acculturation) and for each of the three levels of manufacturing policy. This approach minimizes aggregation and preserves a satisfying degree of realism.

As an example, information was obtained about the use of direct influence for six major manufacturing decisions at Corning and seven at Monsanto. At Corning, the parent had some control over 50 percent of the decisions (three of six), either through absolute centralization, participative centralization, or cooperation. In contrast, Monsanto influenced six of seven decisions (86 percent) in these ways. Evidently the parent's direct influence on major manufacturing decisions of foreign subsidiaries is much greater at Monsanto than it is at Corning.

For standards and procedures, the difference is also very clear. At Monsanto, the direct influence of the parent is much more pervasive than it is at Corning. The dominant pattern at Monsanto is centralization (four out of five decisions), while it is freedom (four out of five) at Corning.

The difference between Corning and Monsanto is much less obvious for the day-to-day manufacturing decisions of the subsidiary. All such decisions at Corning's subsidiaries are made without parent involvement, as are 71 percent (five of seven) of Monsanto's.

Budget processes provide a useful indication of indirect influence through control systems. At Corning the budget is a set of forecasts proposed by the subsidiary and only discussed by the parent. But at Monsanto, the yearly budget is a set of objectives imposed on the subsidiary. In this form of influence, too, Monsanto seems more active than Corning.

For indirect influence through acculturation, the percentage of the foreign subsidiary executives with experience in the parent company was used as a proxy. Four levels of acculturation were somewhat arbitrarily defined:

1. *High-managed acculturation.* More than 30 percent of the subsidiary's top manufacturing executives are local nationals

with more than six months' experience in the parent company.

2. *Low-managed acculturation.* More than 10 percent but less than 30 percent of the subsidiary's top manufacturing executives are local nationals with more than six months' experience in the parent company.

3. *Exported acculturation.* More than 10 percent of the subsidiary's top manufacturing executives are former parent-company executives.

4. *No acculturation.* Less than 10 percent of the subsidiary's top manufacturing executives have parent experience.

Monsanto falls into the category of "no acculturation," while Corning is characterized by "exported acculturation."

To summarize, Monsanto uses a much greater amount of direct influence than Corning on major decisions and on standards and procedures; and Monsanto uses a little more direct influence than Corning on day-to-day decisions. In terms of indirect influence, Monsanto exercises quite a bit more influence than Corning through control systems but somewhat less through acculturation.

This method of comparison has an advantage over a simple assessment of the degree of centralization versus decentralization, because it considers all the methods of influence and introduces the possibility of cooperation. It also differentiates between different levels of manufacturing management. A global picture of the two firms' respective positions can be obtained by aggregating several individual decisions of similar type.

Characteristics of the Firm's Manufacturing Task

The manufacturing task of a firm can be usefully characterized in terms of complexity and uncertainty.

Complexity of task (technology).   It was hypothesized that the more complex the technology, the greater would be the parent's influence. As technology becomes more complex, one would expect the parent to have an increasing competency advantage over the subsidiary. The economies of scale in research, development, engineering, technical abilities, and organization should push the multinational firm to concentrate information processing and decision making at the parent's level.

In manufacturing, the best proxy for complexity is the technology used by the firm. Using Woodward's categories, one can differentiate six types of technologies: small batches, large batches, assembly line, mass production, discontinuous flows, and continuous flows.[10] To simplify, these six categories were grouped into three types:

1. Batches (large and small) ("low technology")
2. Assembly line and mass production ("medium technology")
3. Flows (discontinuous and continuous) ("high technology")

To determine whether technology had any impact on influence, the observations were classified according to these three types of technologies and then examined for patterns of influence characteristic of each type.

Table 12–2 summarizes the results. Technology seems to have an important impact on major manufacturing decisions. All of the firms using batch processes give their manufacturing subsidiaries freedom—either supervised or total—to make major manufacturing decisions. In contrast, the subsidiary has such freedom in only 33 percent of the firms using an assembly line or mass production process and in 19 percent of the firms using a discontinuous or continuous flow process.

Technology also seems to affect the parent's direct influence on standards and procedures, but to a much lesser extent than for major manufacturing decisions. None of the low-technology firms is centralized, 8 percent of the medium-technology firms are, and 24 percent of the high-technology firms are. This result, albeit weak, is consistent with the hypothesis.

Technology does not seem to affect the level of centralization of the day-to-day decisions and seems to have little effect on the use of indirect control systems to influence manufacturing decisions.

The more complex the technology, the greater the parent's direct influence on major manufacturing decisions and perhaps on standards and procedures. Firms with a complex technology tend to impose a manufacturing system on their subsidiaries. The parent itself designs the system and sometimes provides the procedures and standards to operate it. The freedom that the subsidiary then has to make day-to-day decisions is determined by factors other than technology.

Uncertainty of task.   A similar analysis was performed to determine whether firms that confront a very unstable environment—and for which manufacturing consequently is an uncertain task—differ in terms of influence from firms with a stable environment and a certain task. The ability to forecast sales was used as a proxy for uncertainty. Huge forecasting errors mean an uncertain environment, whereas a very high forecasting accuracy suggests a certain environment. The hypothesis was that the more uncertain the environment, the more freedom the parent company would give its foreign subsidiary; i.e., fewer firms would centralize decisions in the parent.

The results, summarized in table 12–3, tend to confirm the hypothesis in two areas of manufacturing policy—major decisions, and standards and procedures. Of the firms with low uncertainty in the manufacturing task, 47 percent centralized major decisions (either absolute or participative) in the parent; 36 percent of those with average uncertainty did so, and only 17 percent of those with high uncertainty. The day-to-day management decisions seemed unaffected by the uncertainty of the manufacturing task.

The results in regard to standards and procedures are ambiguous. One analysis suggests a linear relationship between degree of uncertainty and level of centralization. The parent centralizes decisions concerning standards and procedures for 24 percent of the firms with low uncertainty, for 10 percent of those with average uncertainty, and for none of the firms with high uncertainty. But another analysis suggests a nonlinear relationship. Of the low-uncertainty firms, 53 percent have complete freedom, as compared with only 20 percent of the average-uncertainty firms and 67 percent of the high-uncertainty firms.

Uncertainty also seems to have a nonlinear effect on the use of control systems to exert indirect influence—only 6 percent of the firms with low uncertainty report a high level of such influence, as compared with 44 percent of medium-uncertainty firms and none of the high-uncertainty firms.

An explanation of these somewhat puzzling results may require a distinction between the parent's *desire* and its *ability* to influence the subsidiary. When uncertainty is low, the parent company need only help the subsidiary design its manufacturing system and then may set it free, because the operation of the system is relatively simple and without surprises. The parent company *could* easily centralize but has

TABLE 12–2

EFFECT OF TECHNOLOGY ON LEVEL OF PARENT'S INFLUENCE ON MANUFACTURING DECISIONS OF FOREIGN SUBSIDIARIES, 33 MULTINATIONAL ENTERPRISES

CATEGORY OF DECISIONS	LEVEL OF INFLUENCE	TYPE OF TECHNOLOGY[a] (results expressed as a percentage)			MERGED DATA FOR STATISTICAL TESTS (in absolute numbers)		
		Small/Large Batches	Assembly/Mass Production	Discontinuous/ Continuous Flows	Batches	Other	Probability[b,c]
A. OBTAINED DIRECTLY THROUGH CENTRALIZED SYSTEMS							
Major	Centralization (absolute or participative)	0	42	56 }	0	21	
	Cooperation	0	25	25 }			
	Freedom (supervised or total)	100	33	19	5	7	1.00
	Total	100	100	100			
Standards and procedures	Centralization	0	8	24	0	5	
	Cooperation	60	42	35 }			
	Freedom	40	50	41 }	5	24	1.00
	Total	100	100	100			

Day-to-day					
Centralization	20	36	18	1	7	.66
Cooperation	40	36	35 }	4	21
Freedom	40	27	47
Total	100	100	100

B. OBTAINED INDIRECTLY THROUGH CONTROL SYSTEMS

High	0	27	12
Average	80	36	47
Low	20	36	41
Total	100	100	100
Number of observations for each category	5	11–12	16–17

SOURCE: Author's interviews.

[a] Not all columns total 100 percent, because of rounding.

[b] Probability that the portion of subsidiaries with more freedom or less centralization to make manufacturing decisions in a population of subsidiaries with batch processes is higher than in a population with other manufacturing processes.

[c] The population in note b is the population for which the observations in this study could be considered to be a random sample. Based on prior-posterior analysis using beta distribution. See Robert O. Schlaifer, *Analysis of Decisions Under Uncertainty* (New York: McGraw-Hill, 1967), pp. 11.8–11.16.

TABLE 12–3

EFFECT OF UNCERTAINTY ON LEVEL OF PARENT'S INFLUENCE ON MANUFACTURING DECISIONS OF FOREIGN SUBSIDIARIES, 33 MULTINATIONAL ENTERPRISES

CATEGORY OF DECISIONS	LEVEL OF INFLUENCE	DEGREE OF UNCERTAINTY[a] (results expressed as a percentage)			MERGED DATA FOR STATISTICAL TESTS (in absolute numbers)		
		Low	Average	High	Low and Average	High	Probability[b]
A. OBTAINED DIRECTLY THROUGH CENTRALIZED DECISIONS							
Major	Centralization (absolute or participative)	47	36	17	12	1	
	Cooperation	18	27	33 ⎱	16	5	.92
	Freedom (supervised or total)	35	36	50 ⎰			
	Total	100	100	100			
Standards and procedures	Centralization	24	10	0 ⎱	16	2	
	Cooperation	24	70	33 ⎰	11	4	.89
	Freedom	53	20	67			
	Total	100	100	100			

Day-to-day					
Centralization	22	22	33		.69
Cooperation	28	44	33		2
Freedom	50	33	33 }	6	4
Total	100	100	100	21	

B. OBTAINED INDIRECTLY THROUGH CONTROL SYSTEMS

High	6	44	0
Average	50	33	67
Low	44	22	33
Total	100	100	100

Number of observations for each category	16–18	9–11	6

SOURCE: Author's interviews.

[a] Average error for sales forecast for next year is: Low = 2 percent or less; Average = between 2 and 10 percent; High = 10 percent or more. Not all columns total 100, because of rounding.

[b] Probability that the portion of subsidiaries with more freedom or less centralization to make manufacturing decisions in a population of subsidiaries with high uncertainties is higher than in a population with low or average uncertainties is higher than. See note c, table 12–2.

no incentive to do so. On the other hand, when uncertainty is average, managing the subsidiary becomes a more difficult task, so the parent company tends to exert a higher influence, especially indirectly, through control systems and perhaps directly through standards and procedures; the uncertainty is not so high that it represents an obstacle to the parent's influence. When uncertainty is very high, the parent company probably would like to have a strong influence on the foreign subsidiary, which is difficult to manage, but because it *cannot* achieve such influence it gives up and sets the subsidiary free. The chief point is that very high uncertainty seems to represent a severe obstacle to the execution of high direct influence, or of high indirect influence through control systems.

Two characteristics of the firm's manufacturing task, therefore, seem to have a bearing on influence. The level of the task's technology (or complexity) seems to increase significantly the parent's direct influence on major manufacturing decisions, but it has only a weak effect on direct influence of standards and procedures and forms of influence, and no effect on day-to-day decisions. A high uncertainty of the manufacturing task results in a relatively low level of all forms of influence.

Characteristics of the Firm's Manufacturing System

Two characteristics of the firm's manufacturing system seem to have a bearing on influence: its degree of integration, i.e., the interdependence between the subunits (foreign plants); and its heterogeneity.

Interdependence. The relationship between interdependence and influence seems very strong. Two main types of interdependence were found: physical and nonphysical. The former type is exemplified by cross-shipments of components, semiproducts, or finished products between plants abroad. Nonphysical interdependence may be strategic (the capacity utilization of one plant is related to the capacity utilization of another, or a manufacturing decision taken by one subsidiary might trigger a competitor's move that would affect another subsidiary), or it may relate to marketing (the production of a poor-quality product in a subsidiary may jeopardize an international brand name and thus affect other subsidiaries' performance).

The firms in the sample were grouped into three categories:

1. *Low interdependence.* Foreign subsidiaries are essentially self-contained. Physical linkages are minimal (cross-shipments are no more than 5 percent of the average plant's output); nonphysical linkages between subsidiaries also are minimal.
2. *Average interdependence.* There are some physical linkages (between 5 and 10 percent of the average foreign plant's output) and some nonphysical linkages (either strategic or marketing).
3. *High interdependence.* Cross-shipments are greater than 10 percent of the average plant's output.

As shown in table 12–4, for all three categories of manufacturing decisions—major, standards and procedures, and day-to-day—direct influence increases with interdependence. For example, of eight firms in which interdependence is low, none centralizes major manufacturing decisions, but 56 percent (five out of nine) of the firms in which interdependence is high do so. Similar patterns exist for the other levels of manufacturing policy, and for indirect influence through control systems. These results, of course, are consistent with the hypothesis.

Interdependence between subsidiaries seems thus to trigger a significant increase in the influence the parent company exerts on its subsidiaries, on all levels of manufacturing decisions and through all forms of influence. Interdependence apparently has a more consistent effect on influence than either technology (which has a strong impact, but on major decisions only) or uncertainty (which has a relatively ambiguous impact).

Heterogeneity. It is easier for the parent to exert a strong influence on plants abroad if they are very similar to domestic ones; conversely, influence is more difficult to exert if plants abroad differ in size from the domestic plants or vary in terms of products manufactured, processes used, work-force structure, and vertical integration.

TABLE 12–4

Effect of Interdependence on Level of Parent's Influence on Manufacturing Decisions of Foreign Subsidiaries, 33 Multinational Enterprises

Category of Decisions	Level of Influence	Degree of Interdependence[a] (results expressed as a percentage)			Merged Data for Statistical Tests (in absolute numbers)		
		Low	Average	High	Low	Average and High	Probability[b]
A. OBTAINED DIRECTLY THROUGH CENTRALIZED DECISIONS							
Major	Centralization (absolute or participative)	0	38	56 }	1	20	1.00
	Cooperation	13	38	22 }			
	Freedom (supervised or total)	88	25	22	6	6	
	Total	100	100	100			
Standards and procedures	Centralization	0	13	33 }	0	17	1.00
	Cooperation	0	50	44 }			
	Freedom	100	38	22	8	8	
	Total	100	100	100			

Day-to-day						
Centralization	13	25	44 }	1	19	1.00
Cooperation	0	38	56 }	7	6	
Freedom	88	38	0			
Total	100	100	100			

B. OBTAINED INDIRECTLY THROUGH CONTROL SYSTEMS

High	0	13	33
Average	50	50	44
Low	50	38	22
Total	100	100	100

Number of observations for each category	7–8	16	9–10

SOURCE: Author's interviews.

[a] See text for definitions. Not all columns total 100, because of rounding.

[b] Probability that the portion of subsidiaries with more freedom to make manufacturing decisions in a population of subsidiaries with low interdependence is higher than in a population with average or high interdependence. See note c, table 12–2.

With respect to heterogeneity, three classes of firms can be defined:

1. *Low heterogeneity*. Heterogeneity is low if the output of the biggest foreign plant is at least 50 percent of the output of the biggest domestic plant, and if the foreign plants are similar in at least two of the four factors listed above (products, processes, work-force structure, vertical integration).
2. *Average heterogeneity*. Heterogeneity is average if the biggest foreign plant's output is between 15 percent and 50 percent of the biggest domestic plant, and if the foreign plants are similar in at least two of the four factors listed above.
3. *High heterogeneity*. Heterogeneity is high if the biggest foreign plant has an output less than 15 percent of the biggest domestic plants, and/or if the foreign plants differ in three or more of the factors listed above.

High heterogeneity seems to reduce the parent's direct influence on all three categories of manufacturing decisions, as well as its influence obtained indirectly through control systems. For example, 13 percent (one out of eight) of the high-heterogeneity firms centralized major manufacturing decisions, as compared with 55 and 43 percent of the firms with average or low heterogeneity, respectively. Similar results (although less strong) were found in the other categories of manufacturing decisions and in indirect influence (see table 12–5).

CONCLUSIONS

The influence of a parent company on the manufacturing decisions of its foreign subsidiaries seems to be a function of two characteristics of the task (complexity and uncertainty) and two characteristics of the system (interdependence and heterogeneity).

Two of the variables, interdependence and complexity of technology, have a positive impact on influence. When these "push variables" are high, they encourage the parent to increase its influence on its foreign subsidiaries. The two other variables, uncertainty and heterogeneity, have a negative impact, representing an obstacle to high influence. They could be considered "barrier" variables.

In terms of intensity of impact, interdependence seems the most potent of the four variables. It has a very significant impact on direct influence on all three levels of manufacturing policy and on indirect influence through control systems. Technology also has a strong impact, but primarily on direct influence on major manufacturing decisions.

Uncertainty and heterogeneity have a weaker effect. Their negative impact is most evident when they are high. For firms in the "low" and "average" categories, the effects of differences in those variables are small (heterogeneity) or paradoxical (uncertainty).

These four variables obviously do not exhaust explanations of influence, which may also depend on such general characteristics of the firm as its size or the nationality of the parent company. Characteristics of the peripheral unit, mainly its performance and its distance from the parent, should also affect influence.

The parent company, after all, attempts to maintain a certain level of performance by the subsidiary, and a very poor performance will trigger an increase in influence. Deteriorating performance should thus tend to increase influence. The rapidity of the parent's response will depend on the quality and quantity of the information it receives and the frequency with which it receives it.

Distance between the subsidiary and the parent will represent an obstacle to influence. Alsegg has shown, for example, that influence tends to decrease with geographical distance.[11] But "distance" should be interpreted broadly. Cultural, economic, and demographic differences also presumably represent barriers to influence.

Although these other characteristics of the system and of the subsidiary can also help to explain influence, I would hypothesize that the four variables examined here explain most of the phenomenon. In a multiple regression model with a linear equation, the two variables *technology* and *interdependence* were statistically significant in explaining direct influence (R^2 of 0.5). *Uncertainty* and *heterogeneity* are not linearly related to influence and thus made no additional contribution to the statistical significance of the equation.

This hypothesis was reinforced by the comments of the executives interviewed and by a series of paired comparisons of firms. For example, one of the major differences between W. R. Grace chemicals division and Air Liquide is that Grace's task seems much more

TABLE 12-5

Effect of Heterogeneity on Level of Parent's Influence on Manufacturing Decisions of Foreign Subsidiaries, 33 Multinational Enterprises

Category of Decisions	Level of Influence	Degree of Heterogeneity[a] (results expressed as a percentage)			Merged Data for Statistical Tests (in absolute numbers)		
		Low	Average	High	Low and Average	High	Probability[b]
A. OBTAINED DIRECTLY THROUGH CENTRALIZED DECISIONS							
Major	Centralization (absolute or participative)	43	55	13 }	19	1	
	Cooperation	29	27	0 }			
	Freedom (supervised or total)	29	18	88	6	7	1.00
	Total	100	100	100			
Standards and procedures	Centralization	29	9	0 }	15	2	
	Cooperation	36	45	25 }			
	Freedom	36	45	75	10	6	1.00
	Total	100	100	100			

	Centralization	Cooperation	Freedom	Total			
Day-to-day	29	27	13 }		15	4	.69
	14	55	38 }		10	4	
	57	18	50				
Total	100	100	100				

B. OBTAINED INDIRECTLY THROUGH CONTROL SYSTEMS

High	23	18	0	
Average	46	55	38	
Low	31	27	63	
Total	100	100	100	

Number of observations for each category	13–14	11	8

SOURCE: Author's interviews.
[a]See text for definitions. Not all columns total 100, because of rounding.
[b]Probability that the portion of subsidiaries with more freedom to make manufacturing decisions in a population of subsidiaries with high heterogeneity is higher than a population with low or average heterogeneity. See note c, table 12–2.

uncertain than Air Liquide's—and Grace's foreign subsidiaries enjoy a far higher degree of freedom than Air Liquide's. Similarly, heterogeneity seems much higher in the bicycle division of Peugeot than in the car division—and the influence of the parent is much lower in the bicycle division.

Supported by this battery of indicators (statistical results, qualitative interviews, and comparisons of a sample of firms), the hypotheses seem to warrant a more extensive test on multinational firms (or even in other types of social systems). The present study was primarily concerned with the intensity of vertical influence. It would be useful also to develop and test hypotheses with regard to the intensity of lateral influence. Also, further study is warranted on the balance between formal and informal influence in different firms and on the use of indirect influence—either through control systems or through acculturation. Only then would it be possible to evaluate the impact of specific forms of organization for multinational firms and to make concrete recommendations regarding the design of an influence network in an organization. Ultimately an understanding of influence should be invaluable to further studies of the subjects covered in this volume—choice of technology, channels of technology transfer, and establishment of foreign R&D facilities.

Contributors

The studies in this book were all carried out by students of Professor Raymond Vernon during his years at Harvard Business School. Most of the authors are still actively involved in research on multinational enterprises, pursuing, in many cases, themes begun in the work reported in this volume. Eight of the eleven contributors have been teaching international business: one in Europe, one in Canada, the rest in the United States—New York, Boston, and Washington State. Two are businessmen, but taught for a number of years; another is applying his knowledge of technology transfer in his position in the World Bank.

MICHEL A. AMSALEM is an assistant professor at Columbia University's Graduate School of Business and previously served as the economist in charge of African affairs at the International Finance Corporation, the World Bank Group organization responsible for assistance to the private sector. He is the author of *Technology Choice in Developing Countries* and several articles on technology issues in developing countries. Amsalem's teaching and research interests center on the role of the private sector in economic development and the relations between international firms and developing countries. He is currently doing research on the factors influencing the formation of country and company cartels in raw materials.

HENRI DE BODINAT is the executive vice president at Saatchi and Saatchi Compton in France.

JAMES KEDDIE is currently an advisor to the Nusa Tenggara Barat provincial government in Indonesia. He has also held posts in universities and in the British government, and has served as a consultant to various international agencies, including the International Labour Office in Geneva. His broad interest in industrial development problems, in both developed and developing countries, focuses particularly on the effective diffusion and application of technology in business, with emphasis on meshing technological and marketing factors.

DONALD J. LECRAW is an associate professor of economics and international trade at the School of Business Administration, University of Western Ontario. He has published books and articles on industrial organization, choice of technology, economic development, and international trade.

FARSHAD RAFII is an assistant professor at Boston University's School of Management. He was formerly associated with the Iran Center for Management Studies and has taught in several executive education programs in Iran and the United States. While his interest in the area of international transfer of technology continues, his research has also examined the management of technological innovation and the formulation and communication of manufacturing strategies. He is the coauthor of a recent article entitled "Interfacing Competitive Goals with Manufacturing Strategies."

ROBERT RONSTADT is an associate professor of management and organizational behavior at Babson College and academic head of Babson's entrepreneurial studies program. He is the author of *Research and Development Abroad by U.S. Multinationals* and has recently published several articles in the area of entrepreneurship.

ROBERT STOBAUGH is the Charles E. Wilson Professor of Business Administration at Harvard Business School. He spent eighteen years in the oil and petrochemical industries in the United States and abroad before earning his doctorate at Harvard Business School. Author, coauthor, or coeditor of nine books in three subject areas— international business, technology and production, and energy—he is a past president of the Academy of International Business. Stobaugh's best-known book, of which he was a principal author and coeditor, is *Energy Future: A Report of the Energy Project at the Harvard Business*

School. His most recent research has been on the innovation of new products and processes and the subsequent spread of technology.

PIERO TELESIO is a business consultant specializing in manufacturing policy and currently teaches at the Management Education Institute at Arthur D. Little, Inc. He has also taught at the Boston University School of Management and the Fletcher School of Law and Diplomacy, Tufts University. His publications include *Technology Licensing and Multinational Enterprises* and a recent *Harvard Business Review* article (with Stobaugh) entitled "Match Manufacturing Policies and Product Strategy."

LOUIS T. WELLS, JR., is Herbert F. Johnson Professor of International Management at Harvard Business School. He has been doing research on multinational firms based in developing countries and on government structures and procedures to negotiate with foreign investors. His recent publications include *Third World Multinationals* and an article on "Bargaining Power of Multinationals and Host Governments."

DAVID WILLIAMS was formerly an industrial advisor to several countries, including Tanzania, with the Harvard Institute for International Development. He is now chief of the Industrial Development and Finance Division of the South Asia Projects Department of the World Bank. In this capacity he deals with a wide range of industrial issues, including the problems of state enterprises, in borrowing countries.

WAYNE A. YEOMAN has been senior vice president—finance at Eastern Airlines, Inc., since 1976. He was formerly a member of the faculty of the United States Air Force Academy.

Notes

CHAPTER ONE NOTES

1. For surveys of the conventional literature and of other recent empirical work, see David Morawetz, *Twenty-Five Years of Economic Development, 1950–1975* (Baltimore: Johns Hopkins University Press, 1978); and Frances Stewart, *Technology and Underdevelopment* (Boulder, Col.: Westview Press, 1977).

2. See, for example, Richard Eckaus, "The Factor Proportions Problem in Undeveloped Areas," *American Economic Review,* Vol. 45 (September 1955).

3. The ideas in Oliver Williamson's *Markets and Hierarchies: Analysis and Antitrust Implications: A Study in the Economics of Internal Organization* (New York: Free Press, 1975) are only recently providing a framework for integrating into economic models the kind of work done in these studies.

4. W. P. Strassmann, *Technological Change and Economic Development: The Manufacturing Experience of Mexico and Puerto Rico* (Ithaca: Cornell University Press, 1968).

5. Particularly good, for example, are the empirical studies by Howard Pack: "The Substitution of Labour for Capital in Kenyan Manufacturing," *Economic Journal,* Vol. 86 (March 1977); "The Choice of Technique in Cotton Textiles," mimeographed, January 1974; "Employment and Productivity in Kenyan Manufacturing," *East Africa Economic Review,* Vol. 4 (December 1972); by C. Peter Timmer: "Using a Probabilistic Frontier Production Function to Measure Technical Efficiency," *Journal of Political Economy,* Vol. 79 (July–August 1971); and by Gus Ranis: "The Role of the Industrial Sector in Korea's Transition to Economic Maturity," Center Discussion Paper No. 25 (New Haven: Yale University Economic Growth Center, October 1971).

6. W. D. Morrison, vice president of Hooker Chemical Corporation, quoted in *Oil, Paint and Drug Reporter* (now *Chemical Marketing Reporter*), March 20, 1967, p. 32.

7. P. P. Gabriel, "The Transfer of Organization and Technological Resources to Less Developed Countries Through Management Contracts Between Local Public Enterprises and Foreign Private Firms," D.B.A. dissertation, Harvard Business School, 1965, from which was developed his book, *International Transfer of Corporate Skills: Management Contracts in Less Developed Countries* (Boston: Harvard Business School, Division of Research, 1967).

8. These findings are consistent with the subsequently developed argument that ownership is preferred over licensing when there are market imperfections, such as lack of a futures market in the technology, the ability to discriminate in the product pricing, bilateral concentration, inequality between buyer and seller with respect to knowledge of the technology, public-good characteristics, and the ability to manipulate transfer prices for technology. See P. J. Buckley and Mark Casson, *The Future of Multinational Enterprise* (New York: Holmes and Meier, 1976), Chapter 2.

9. Oliver Williamson, *op. cit.* See also Buckley and Casson, *op. cit.,* and Louis T. Wells, Jr., *Third World Multinationals* (Cambridge, Mass.: MIT Press, 1983).

297

CHAPTER TWO NOTES

1. See, in particular, M. L. Olson, Jr., *The Economics of Wartime Shortage: A History of British Food Suppliers in the Napoleonic War and in World Wars I and II* (Durham, N.C.: Duke University Press, 1963); "The Economics of Target Selection," *The Royal United Service Institution Journal*, Vol. 107 (November 1962), pp. 308–314; and "American Materials Policy and the Physiocratic Fallacy," *Orbis*, Vol. 6 (Winter 1963), pp. 670–688.

2. See G. K. Boon, *Economic Choice of Human and Physical Factors in Production* (Amsterdam: North Holland Publishing Company, 1964); *Optimal Capital Intensity in Metal-Chipping Processes*, Progress Report No. 1 (Stanford, Calif.: Stanford University Institute of Engineering-Economic Systems, October 1965); and *Optimal Technology in Machine-Chipping Tools*, Progress Report No. 2 (Stanford, Calif.: Stanford University Institute of Engineering-Economic Systems, May 1966). See also E. Staley and R. Morse, *Modern Small Industry for Developing Countries* (New York: McGraw-Hill, 1965).

3. For an argument supporting this finding, see A. O. Hirschman, *The Strategy of Economic Development* (New Haven: Yale University Press, 1958), pp. 139–155.

4. See G. K. Boon, *Optimal Capital Intensity in Metal-Chipping Processes, op. cit.*, p. 30.

5. Cross-elasticity of demand measures the tendency of buyers to substitute one product for another when the price of one of the products changes. A low cross-elasticity indicates poor substitutes (high degree of product differentiation), and a high cross-elasticity indicates close substitutes. As the individual firm alters the pattern of differentiation of its products, it alters the cross-elasticity of demand for those products and, in turn, the shape of its individual demand curve.

6. Price theory contends, of course, that the demand of firms for a particular factor is related solely to the marginal revenue product of that factor. Theoretically, profit maximizers hire units of productive factors (for the case at issue, technicians and funds to develop alternative production processes) up to the point at which the marginal outlay in purchasing the factors is equal to the marginal value contributed by the factors. A failure even to begin the effort to develop alternative processes is clearly not a rational response, but the field research indicates that it is a fact of business life. Melvyn Copen noted the same phenomenon during his study of the operations of U.S. drug companies in India. Copen reports: "It is worth noting that all of the U.S. top executives primarily concerned themselves with marketing to the neglect of other functions—even those executives who were not originally marketing men." See M. R. Copen, "The Management of U.S. Manufacturing Subsidiaries in a Developing Nation: India," D.B.A. dissertation, Harvard Business School, 1967, p. 89.

7. While the response described for Firm A seems to be the typical one, an alternative course of action could be rational for certain multinational companies. With the labor and capital content of product selling price low, and with product quality easy to maintain, corporate executives may be quite indifferent to the kind of production techniques employed abroad. Under these conditions, the selection of manufacturing methods could be a matter for local managers to resolve, and thus capital-labor ratios might vary widely across these systems.

8. M. R. Copen, *op. cit.*, p. 50.

9. *Ibid.*, pp. 110–115.

10. J. S. Bain, *Barriers to New Competition* (Cambridge, Mass.: Harvard University Press, 1956), pp. 169–170.

11. *Ibid.*, p. 176.

12. Entry problems can be further reduced by subcontracting for components. In a recent year (in the 1960s), B_6 subcontracted 48 percent of the production work on

the elevator units assembled in its Indian plant. If competent subcontractors could be found in India, they could probably be found in many other areas of the world as well.

13. An analysis of the structure of the appliance industry abroad confirms this position. The data in U.S. Department of Commerce, *Major Household Appliances—Production, Consumption, Trade—Selected Foreign Countries,* (Washington: G.P.O., September 1960) identify important patterns in the industry. The report makes it clear that by 1960 the appliance industry was highly developed in many countries abroad. The technology of appliance manufacture is relatively simple, and unskilled labor is easily trained to handle the various production tasks. There are many competing firms.

CHAPTER THREE NOTES

1. See, for example, A. E. Kahn, "Investment Criteria in Development," *Quarterly Journal of Economics,* Vol. 65 (February 1951), pp. 38–51; H. B. Chenery, "The Application of Investment Criteria," *Quarterly Journal of Economics,* Vol. 67 (February 1953), pp. 76–96; W. Galenson and H. Leibenstein, "Investment Criteria, Productivity and Economic Development," *Quarterly Journal of Economics,* Vol. 69 (August 1955), pp. 343–370; R. S. Eckaus, "The Factor Proportions Problem in Underdeveloped Areas," *American Economic Review,* Vol. 45 (September 1955), pp. 539–565; O. Eckstein, "Investment Criteria for Economic Development and the Theory of Intertemporal Welfare Economics," *Quarterly Journal of Economics,* Vol. 71 (February 1957), pp. 56–85; A. K. Sen, "Some Notes on the Choice of Capital Intensity in Development Planning," *Quarterly Journal of Economics,* Vol. 71 (November 1957), pp. 561–584.

2. See, for example, G. K. Boon, *Economic Choice of Human and Physical Factors in Production* (Amsterdam: North Holland Publishing Company, 1964); United Nations, Department of Economic and Social Affairs, *Industrialization and Productivity,* Bulletin 3 (March 1960); A. K. Sen, *Choice of Techniques* (Oxford: Blackwells, 1960).

3. See, for example, L. Reynolds, *Wages, Productivity and Industrialization in Puerto Rico* (Homewood, Ill.: Richard D. Irwin, Inc., 1965); M. Nerlove, "Recent Empirical Studies of the CES and Related Production Functions," in *The Theory and Empirical Analysis of Production,* Studies in Income and Wealth, Vol. 21 (New York: N.B.E.R., 1967); I. Sakong, *Factor Market Price Distortions and Choice of Production Techniques in Developing Countries,* Ph.D. dissertation, Graduate School of Business Administration, UCLA, 1969; Z. Griliches, "Capital-Skill Complementarity," *Review of Economics and Statistics,* Vol. 51 (November 1969), pp. 465–468; R. H. Mason and I. Sakong, "Level of Economic Development and Capital Labor Ratios in Manufacturing," *Review of Economics and Statistics,* Vol. 53 (May 1971), pp. 176–178. For a summary of such studies, see H. J. Bruton, "The Elasticity of Substitution in Developing Countries," Research Memorandum No. 45 (Williamstown, Mass.: Williams College Center for Development Economics, April 1972).

4. For a particularly useful study in this classification, see W. P. Strassmann, *Technological Change and Economic Development: The Manufacturing Experience of Mexico and Puerto Rico* (Ithaca: Cornell University Press, 1968). See also R. H. Mason, "The Transfer of Technology and the Factor Proportions Problem: The Philippines and Mexico," United Nations Institute for Training and Research, *UNITAR Research Report No. 10,* (n.d.).

5. Betjaks are the pedicabs in common use in Indonesia. Betjak tires, while they are the same sizes as bicycle tires, are generally made for much heavier use.
6. If two distinct technologies were used by the same firm for a product, or if two products included in this study were manufactured by a firm, each product-technology combination was counted as a plant. Two processes for the same product in the same firm were encountered in three cases.
7. This is not a result of uneven distribution of foreign investors across industries.
8. Appendix C describes the processes analyzed to arrive at the investment per worker saved. The range of cost estimates for alternative machines varied greatly from firm to firm, indicating that alternative technologies had rarely been looked into carefully.
9. All calculations were made for a constant output. For example, if to produce 1,000 items per day, two machines costing $500 and requiring two workers each were required at the labor-intensive level, and one machine costing $1,500 and using two workers was required at the intermediate level, the investment per worker saved would be $250.
10. The problems of variations in plant size and in period of choice of equipment were relatively minor in this study. See Appendix C.
11. Note that they were replacing intermediate equipment, for which most of the costs would not be recovered. Thus, their investment in eliminating a worker would generally be higher than those calculated in the previous section.
12. Kretek cigarettes contain cloves, in contrast to the so-called white cigarettes.
13. A similar result was found for plywood manufacture in Korea. See G. Ranis, "The Role of the Industrial Sector in Korea's Transition to Economic Maturity," Center Discussion Paper No. 25 (New Haven: Yale University Economic Growth Center, October 1971). In fact, economic historians have claimed that the substitution of capital for labor in the industrialization of Western countries resulted in higher raw material usage.
14. The third firm was in an industry that was protected and for which domestic capacity was far below domestic demand.
15. Chapter two of this book reports the results of a study in which Wayne Yeoman pointed out that a U.S. firm was more likely to depart from its U.S. technology in low-wage countries if the primary basis of competition was price.
16. The Yeoman study, referred to in the preceding footnote, indicated that American firms adjust their technology more, in low-wage countries, for light manufacturing than for heavy manufacturing. The greater adjustment for light manufacturing probably reflects the existence of a wider range of known, feasible technologies for light industries.
17. With the many steps and the several alternative processes involved in the making of batteries, battery manufacture provided more of a continuum of technological alternatives than did the other industries.

CHAPTER FOUR NOTES

1. See p. 58 of previous chapter.
2. W. P. Strassmann, *Technological Change and Economic Development: The Manufacturing Experience of Mexico and Puerto Rico* (Ithaca: Cornell University Press, 1968), p. 154.
3. See p. 63 of previous chapter.
4. W. A. Yeoman, "Selection of Production Processes for the Manufacturing Subsidiaries of U.S.-Based Multinational Corporations," D.B.A. dissertation, Harvard Business School, 1968, Chapter 4.

CHAPTER FIVE NOTES

1. R. B. Stobaugh et al., *Nine Investments Abroad and Their Impact at Home* (Boston: Harvard Business School, Division of Research, distributed through Harvard University Press, 1976), pp. 201–202; S. A. Morley and G. W. Smith, "Limited Search and the Technology Choices of Multinational Firms in Brazil," *Quarterly Journal of Economics*, Vol. 91 (May 1977), pp. 263–287; J. Bergsman, "Commercial Policy, Allocative Efficiency and 'X-Efficiency,'" *Quarterly Journal of Economics*, Vol. 96 (August 1974), pp. 409–433; L. H. White, "Appropriate Technology, X-Inefficiency and a Competitive Environment: Some Evidence from Pakistan," *Quarterly Journal of Economics*, Vol. 90 (November 1976), pp. 575–589. White compared the capital intensity of firms in Pakistan with those in the United States. He found that firms in concentrated industries in Pakistan chose a technology that was more capital intensive, relative to those in the same industry in the United States, than did firms in more competitive industries.

2. See R. Sato, "Homothetic and Non-Homothetic CES Production Functions," *American Economic Review*, Vol. 67 (September 1977), pp. 559–569.

3. K. Arrow, H. Chenery, B. Minhas, and R. Solow, "Capital-Labor Substitution and Economic Efficiency," *Review of Economics and Statistics*, Vol. 43 (August 1974), pp. 225–250.

4. See in particular, D. Morawetz, "Elasticities of Substitution in Industry: What Do We Learn from Econometric Estimates?" *World Development*, Vol. 4 (January 1976), pp. 11–15. Morawetz found no correlation between different researchers' estimates of the ranking of industries by their elasticity of substitution. This literature is expanding daily. See E. R. Berndt, "Reconciling Alternative Estimates of the Elasticity of Substitution," *Review of Economics and Statistics*, Vol. 59 (February 1976), pp. 59–68; D. A. Fitchett, "Capital-Labor Substitution in the Manufacturing Sector of Panama," *Economic Development and Cultural Change*, Vol. 24 (April 1976), pp. 577–592; D. J. C. Forsyth, "Appropriate Technology in Sugar Manufacturing," *World Development*, Vol. 5 (March 1977), pp. 189–202; K. L. Gupta, "Factor Prices, Expectation and Demand for Labor," *Econometrica*, Vol. 43 (July 1975), pp. 757–770; L. Hoffman and B. Weber, "Economies of Scale, Factor Intensities and Substitution: Micro Estimates for Malaysian Manufacturing Industries," *Review of World Economics*, Vol. 112 (January 1976), pp. 111–135; J. M. Kearl, "Aggregate Production Functions: Some CES Experiments," *Review of Economic Studies*, Vol. 44 (June 1977), pp. 305–320; A. A. Kintis, "Specification of the Elasticity of Substitution Function Within Cost Minimization CES Production Function Framework," *Économie Appliquée*, Vol. 29 (January 1976), pp. 33–48; T. P. Lianos, "Capital-Labor Substitution in a Developing Country," *European Economic Review*, Vol. 6 (April 1975), pp. 129–141; E. J. Nwosu, "Some Problems of 'Appropriate' Technology and Technology Transfer," *Developing Economies*, Vol. 13 (March 1975), pp. 82–93; M. Roemer, G. M. Tidrick, and D. Williams, "The Range of Strategic Choice in Tanzanian Industry," *Journal of Development Economics*, Vol. 3 (September 1976), pp. 257–275; K. W. Roskamp, "Labor Productivity and the Elasticity of Factor Substitution in West German Industries," *Review of Economics and Statistics*, Vol. 59 (August 1977), pp. 366–371; J. Schaafsma, "Capital-Labor Substitution and the Employment Function in Manufacturing: A Model Applied to 1949–1972 Canadian Data," *Quarterly Review of Economics and Business*, Vol. 17 (Autumn 1977), pp. 32–42; J. Schaafsma, "On Estimating the Time Structure of Capital-Labor Substitution in the Manufacturing Sector," *Southern Economic Journal*, Vol. 44 (April 1978), pp. 740–751; and H. H. Tsang and J. J. Perskey, "On the Empirical Content of CES Production Function," *Economic Record*, Vol. 51 (December 1975), pp. 539–548.

5. Many other production functions have been suggested. See E. R. Berndt and L. R. Christensen, "The Translog Function and the Substitution of Equipment, Struc-

tures and Labor in U.S. Manufacturing 1929–68," *Journal of Econometrics,* Vol. 1 (March 1973), pp. 81–114; L. R. Christensen, D. W. Jorgenson, and L. J. Lau, "Transcendental Logarithmic Production Frontiers," *Review of Economics and Statistics,* Vol. 55 (February 1973); R. Fare and R. W. Shephard, "Ray Homothetic Production Functions," *Econometrica,* Vol. 45 (January 1977), pp. 133–146; J. H. Gapinski and T. K. Kumar, "Embodiment, Putty-Clay and Misspecification of the Directly Estimated CES," *International Economic Review,* Vol. 17 (June 1976), pp. 472–438; G. Hanoch, "Cresh Production Functions," *Econometrica,* Vol. 29 (September 1971), pp. 695–712; A C. Harvey, "Discrimination Between CES and VES Production Function," *Annals of Economic and Social Measures,* Vol. 6 (Fall 1977), pp. 463–471; K. R. Kadiyala, "Production Functions and Elasticity of Substitution," *Southern Economic Journal,* Vol. 38 (January 1972), pp. 281–284; Y. Lu and L. B. Fletcher, "A Generalization of the CES Production Function," *Review of Economics and Statistics,* Vol. 50 (November 1968), pp. 449–452; N. S. Revankar, "A Class of Variable Elasticity of Substitution Production Functions," *Econometrica,* Vol. 39 (January 1971), pp. 61–71; R. Sato, "Homothetic and Non-Homothetic CES Production Functions," *American Economic Review,* Vol. 67 (September 1977), pp. 559–569; and R. Sato, "The Most General Class of CES Functions," *Econometrica,* Vol. 43 (September–November 1975), pp. 999–1003.

6. See G. C. Winston, "Factor Substitution, Ex Ante and Ex Post," *Journal of Development Economics,* Vol. 1 (September 1974), pp. 145–163, for the problems introduced by different and changing levels of capacity utilization.

7. See H. Tsurumi and Y. Tsurumi, "A Bayesian Estimation of Macro and Micro CES Production Functions," *Journal of Econometrics,* Vol. 4 (February 1976), pp. 1–25; T. K. Kumar and J. H. Gapinski, "Nonlinear Estimation of the CES Production Function: Sampling Distributions and Tests in Small Samples," *Southern Journal of Economics,* Vol. 41 (October 1974), pp. 258–266; and T. K. Kumar and J. H. Gapinski, "Nonlinear Estimation of the CES Production Parameters: A Monte Carlo Study," *Review of Economics and Statistics,* Vol. 56 (November 1974), pp. 563–567.

8. D. Morawetz, "Employment Implications of Industrialization in Developing Countries: A Survey," *Economic Journal,* Vol. 84 (September 1974), pp. 491–542.

9. H. A. Simon, "Theories of Decision Making in Economics and Behavioral Science," *American Economic Review,* Vol. 49 (June 1959), pp. 253–283.

10. H. Leibenstein, *Beyond Economic Man: A New Foundation for Microeconomics* (Cambridge, Mass.: Harvard University Press, 1976).

11. R. A. McCain, "Competition, Information, Redundancy: X-Efficiency and the Cybernetics of the Firm," *Kyklos,* Vol. 28 (1975), pp. 268–308.

12. See H. Leibenstein, *Beyond Economic Man.* G. J. Stigler, "The Xistence of X-Efficiency," *American Economic Review,* Vol. 66 (March 1976), pp. 213–216, disagreed with the entire concept of X-Inefficiency, but Leibenstein answered his charges in "X-Inefficiency Xists—Reply to an Xorcist," *American Economic Review,* Vol. 68 (March 1978), pp. 203–211.

13. See Chapter 3.

14. See Chapter 4.

15. See D. J. Lecraw, "Choice of Technology in Low-Wage Countries," Ph.D. dissertation, Harvard University, 1976. Because the data were collected under a promise of confidentiality, the industries in which the firms operated are identified at the two-digit level in the tables.

16. A smaller range of efficient technologies would probably be found in heavy industries characterized by generally higher capital-labor ratios and continuous flow technologies.

17. Projected capital-labor ratio/actual capital-labor ratio was 0.975 (standard deviation 0.038) for equipment, 0.81 (0.13) for buildings, and 0.93 (0.10) for land.

18. M. S. Feldstein, "Alternative Methods of Estimating a CES Production Function for Britain," *Economica*, Vol. 34 (November 1967), pp. 384-394.

19. See G. C. Winston, *op. cit.*, and G. C. Winston, "The Theory of Capital Utilization and Idleness," *Journal of Economic Literature*, Vol. 12 (December 1975), pp. 1301-1320. I have also explored the factors that influence the level of capacity utilization in D. J. Lecraw, "Determinants of Capacity Utilization by Firms in Low-Wage Countries," *Journal of Development Economics*, Vol. 5 (June 1978), pp. 139-153.

20. See J. Johnston, *Econometric Methods* (New York: McGraw-Hill, 1972), p. 281.

21. See D. J. Aigner and S. I. Chu, "On Estimating the Industry Production Function," *American Economic Review*, Vol. 63 (June 1973), pp. 826-839; D. J. Aigner, C. A. K. Lovell, and P. Schmidt, "Formulation and Estimation of Stochastic Frontier Production Function Models," *Journal of Econometrics*, Vol. 6 (July 1977), pp. 21-37; D. J. Aigner, T. Amemiya, and D. J. Poirier, "On the Estimation of Production Frontiers: Maximum Likelihood Estimation of the Parameters of a Discontinuous Density Function," *International Economic Review*, Vol. 17 (June 1976), pp. 377-396; T. Amemiya, "Regression Analysis When the Dependent Variable is Truncated Normal," *Econometrica*, Vol. 41 (November 1973), pp. 997-1016; F. R. Forsund and E. S. Jansen, "On Estimating Average and Best Practice Homothetic Production Functions Via Cost Functions," *International Economic Review*, Vol. 18 (June 1977), pp. 463-476; D. Leech, "Testing the Error Specification in Nonlinear Regression," *Econometrica*, Vol. 43 (July 1975), pp. 719-725; W. Meeusen and J. V. D. Broeck, "Efficiency Estimation from Cobb-Douglas Production Functions with Composed Error," *International Economic Review*, Vol. 18 (June 1977), pp. 435-444; J. Richmond, "Estimating the Efficiency of Production," *International Economic Review*, Vol. 15 (June 1974), pp. 515-528; and C. P. Timmer, "Using a Probabilistic Frontier Production Function to Measure Technical Efficiency," *Journal of Political Economy*, Vol. 79 (July-August 1971), pp. 776-794.

22. M. J. Farrell, "The Measurement of Productive Efficiency," *Journal of the Royal Statistical Society*, Series A, Vol. 120 (1957), Part 3.

23. Farrell defined a technology as technically efficient "if it uses a smaller input combination for a given amount of product" than any other technology. See P. Meller, "Efficiency Frontiers for Industrial Establishments of Different Sizes," *National Bureau of Economic Research Occasional Papers*, No. 3 (Summer 1976), pp. 397-467.

24. D. J. Aigner, C. A. K. Lovell, and P. Schmidt, *op. cit.* Timmer's method (and the one used here) is "essentially arbitrary; lacking explicit economic or statistical justification." In partial defense, it should once again be emphasized that the results on the determinants of the deviations from low-cost technology were extremely robust for alternative specifications of the production functions and methods of estimations.

25. Meller, *op. cit.*

26. See D. J. Lecraw, "Choice of Technology in Low-Wage Countries," p. 97, Table 4-2.

27. See D. J. Lecraw, "Direct Investment by Firms from LDCs," *Oxford Economic Papers*, Vol. 29 (November 1977), pp. 442-457.

28. Since Thailand's is a small, open economy, imports as a percentage of consumption might also be used. This measure was not used for two reasons. First, the level of effective tariffs on imports was high for the industries in the study, so that import penetration was not a good measure of competitive price pressure. Second, accurate data on imports and outputs at the five-digit level were not available on a yearly basis.

29. R. A. McCain, *op. cit.*

30. See Y. W. Rhee and L. E. Westphal, "A Micro Econometric Investigation of Choice of Technology," *Journal of Development Economics,* Vol. 4 (September 1977), pp. 205-237. They found that government incentive grants for exporters led to a choice of more capital-intensive technologies.

31. It is noteworthy that the correlation should emerge in this analysis since the very choice of inappropriate technology increases a firm's costs and reduces its profits. Thus, otherwise high-profit firms might have been pushed into the low-projected-profit category by such a decision. To eliminate the impact of technology choice on the projected level of profits, each firm's profits were restated as if it had chosen its technology to minimize costs; with this adjustment, the influence of expected profits on technology choice appeared even more important.

32. The conclusions are consistent with those of H. Pack, "The Substitution of Labour for Capital in Kenyan Manufacturing," *Economic Journal,* Vol. 86 (March 1977), pp. 45-58, and of R. F. Solomon and D. J. C. Forsyth, "Substitution of Labour for Capital in the Foreign Sector: Some Further Evidence," *Economic Journal,* Vol. 87 (June 1977), pp. 283-289. Leipziger concluded that foreign firms in India had greater access to information and hence chose more technically efficient techniques. See D. M. Leipziger, "Production Characteristics in Foreign Enclave and Domestic Manufacturing: The Case of India," *World Development,* Vol. 4 (April 1976), pp. 321-325. Cohen concluded, however, that there was no difference between foreign-owned and domestically owned firms in Korea. See B. I. Cohen, "Comparative Behavior of Foreign and Domestic Export Firms in a Developing Economy," *Review of Economics and Statistics,* Vol. 55 (May 1973), pp. 190-197. Hellinger and Hellinger concluded that multinational firms have generally done little to introduce appropriate technologies in LDCs. See D. Hellinger and S. Hellinger, "The Job Crisis in Latin America: A Role for Multinational Corporations in Introducing More Labour-Intensive Technologies," *World Development,* Vol. 3 (June 1975), pp. 399-410. In Thailand, those firms whose owners were based in other LDCs used the most appropriate technology. See D. J. Lecraw, "Direct Investment by Firms from LDCs."

33. H. Pack, "The Substitution of Labour for Capital in Kenyan Manufacturing."

CHAPTER SIX NOTES

1. These results are not surprising for the textile industry, for which several studies have reached the same conclusion. See, for example, H. Pack, "The Choice of Technique in Cotton Textiles," mimeographed, January 1974; J. Pickett and R. Robson, "Technology and Employment in the Production of Cotton Cloth" (University of Strathclyde, Scotland, n.d.); "Technological and Economic Aspects of Establishing Textile Industries in Developing Countries" (Vienna, United Nations, UNIDO, 1967). They are more unexpected for the pulp and paper industry, which is classified as a chemical process industry. While a high level of technological fixity in chemical process industries is often assumed (see, for example, J. Keddie, "Adoption of Production Technique by Industrial Firms in Indonesia," Ph.D. thesis, Harvard University, 1975, and G. K. Boon, *Economic Choice of Human and Physical Factors in Production* [Amsterdam: North Holland, 1964]), very few studies have in fact been done on the scope for capital-labor substitution in these industries.

2. M. A. Amsalem, *Technology Choice in Developing Countries,* Table 2-1, pp. 36-37. Hand looms and multiphase looms not included.

3. See, for example, W. Baer and M. Herve, "Employment and Industrialization in Developing Countries," *Quarterly Journal of Economics,* Vol. 80 (February 1966), pp. 88-107; W. P. Strassmann, *Technological Choice and Economic Development:*

The Manufacturing Experience of Mexico and Puerto Rico (Ithaca: Cornell University Press, 1968).

4. M. A. Amsalem, *Technology Choice in Developing Countries,* pp. 45–46.

5. Pack studied Kenyan manufacturing firms, and pointed to what he called "the substitution between labor and buildings." However, since the cost of buildings was not included in the capital cost of a technology alternative, he did not reach conclusions as to the impact of this higher building requirement of labor-intensive technologies on their competitiveness. H. Pack, "Employment and Productivity in Kenyan Manufacturing," *East Africa Economic Review,* Vol. 4 (December 1972), pp. 29–52.

6. Several studies point to the unavailability of information or to a limited search for information as an explanation for the adoption of capital-intensive technologies. L. T. Wells, Jr., "Economic Man and Engineering Man," Chapter 3 of this volume for Indonesia; S. A. Morley and G. W. Smith, "Managerial Discretion and the Choice of Technology by Multinational Firms in Brazil," Paper No. 56 (Houston: Rice University Program of Development Studies, Fall 1974) for Brazil; and R. H. Mason, *Promoting Economic Development* (Claremont: Claremont College Press, 1955) for the Philippines and Mexico, found that foreign firms had a high propensity to choose equipment from their home countries. This was confirmed by R. B. Stobaugh et al., *Nine Investments Abroad and Their Impact at Home* (Boston: Harvard Business School, Division of Research, 1976), a study of the impact on the United States of foreign investment by U.S. firms. Wells, Morley, and Smith found a strong link between the origin of the equipment and its capital intensity, with imported equipment being more capital intensive than local equipment. However, all these studies investigated industries in which locally manufactured equipment was available. In this respect, this study contrasts sharply with its predecessors, since locally manufactured equipment was not available for the operations studied here. The conclusion is that, instead of purchasing equipment in other developing countries, locally owned firms turn to developed countries, as do foreign subsidiaries. See W. A. Chudson and L. T. Wells, Jr., "The Acquisition of Proprietary Technology by Developing Countries from Multinational Enterprises: A Review of Issues and Policies" (New York: United Nations Department of Economic and Social Affairs, 1974).

7. For an analysis of the firm's motivations, see R. B. Stobaugh et al., *Nine Investments Abroad, op. cit.*

8. The same conclusions were reached by D. Williams, "National Planning and the Choice of Technology: The Case of Textiles in Tanzania," D.B.A. dissertation, Harvard Business School, 1976.

CHAPTER SEVEN NOTES

1. D. Williams, "National Planning and the Choice of Technology: The Case of Textiles in Tanzania," D.B.A. dissertation, Harvard Business School, 1976. Chapter 3.

2. Act No. 69 of 1964 on the National Development Corporation (Section 6), Dar es Salaam: Government Printer.

3. Julius K. Nyerere, Speech to the Board of NDC, January 15, 1965, quoted by Knud Erik Svendsen, "Decision-Making in the National Development Corporation," in Lionel Cliffe and John S. Saul, eds., *Socialism in Tanzania,* Vol. II, Policy (Dar es Salaam: East African Publishing House, 1973).

4. See J. Rweyemamu, *Underdevelopment and Industrialization in Tanzania: A Study of Perverse Capitalist Industrial Development* (Nairobi: Oxford University Press, 1974), p. 124.

5. National Development Corporation, Dar es Salaam, 1968, Fourth Annual Report, p. 58.
6. See A. Coulson, "The Fertilizer Factory," mimeographed. Paper presented to the Conference of the East African Agricultural Economic Society, Dar es Salaam, June 1973.
7. *Ibid.*
8. *Ibid.*
9. R. Vernon, "International Investment and International Trade in the Product Cycle," *Quarterly Journal of Economics,* Vol. 80 (May 1966), pp. 190–207.
10. See Barher, Bhagavas, Collande, and Wield, "Notes on Tanzanian Industries," an unpublished survey, University of Dar es Salaam (n.d.).
11. These consisted of six operating firms: Blanket Manufacturers Ltd., Mwanza Textile Ltd., Friendship Textile Mills Ltd., Kilimanjaro Textile Mills Ltd., National Textile Industries Corporation Ltd., and Tanzania Bag Corporation Ltd.
12. For instance, "Operations Department," "Planning and Finance Department," "Research and Development Department," and so forth, each department under a "director."
13. Both these arguments for expansion of capacity were explored in detail in D. Williams, "Development of the Textile Industry in Tanzania 1975–1995," mimeographed (Dar es Salaam: Devplan, August 1973).
14. See Appendix IV to EEC paper of January 1974, mimeographed, confidential (Dar es Salaam: Government of Tanzania).
15. See National Textile Corporation, "Three Year Investment and Development Plan—1974 to 1976," mimeographed (Dar es Salaam, March 1974).
16. The TAZARA Railroad was nearing completion in Tanzania and the Chinese were seeking other aid projects; their interest was directed toward the development of iron and steel in southern Tanzania.
17. See International Bank for Reconstruction and Development, *Tanzania: Industrial and Mining Sector Survey,* Volumes I, II, and III (Washington, D.C.: IBRD, March 1975).
18. The writer accompanied the textile analyst on all his visits and was seconded by Devplan to provide data developed in previous Devplan investigations of the industry.
19. See National Textile Corporation, "Three-Year Plan," *op. cit.*
20. For an interesting analysis of organizational process in aid agencies, see Judith Tendler, *Inside Foreign Aid* (Baltimore: The Johns Hopkins University Press, 1975). Tendler demonstrates clearly that organizational processes in the U.S. Agency for International Development led to the neglect of alternative suppliers, technologies, and processes when the agency was involved in projects in Brazil.
21. A former general manager of NDC expressed the opinion that discussions that took place during a visit to China by President Nyerere (circa 1964) were the starting point for a textile project and an agricultural implements project. He was not, however, a member of the Tanzanian delegation.
22. To illustrate the inflexibility of the Chinese technicians, one interviewee told how he and others made an unsuccessful attempt to alter the design of some workers' apartments being built for one of the two projects. The design, from China, involved a layout that placed the toilet next to (and off) the dining room. The Chinese were unmoved by Tanzanian protestations that this was not desirable and not customary in Tanzania. The units were built as designed.
23. See National Development Corporation, "Three Year Investment and Development Plan," (1974 to 1976) mimeographed (Dar es Salaam, 1974).
24. See National Development Corporation, "Morogoro Complex," feasibility study by the Investment Advisory Centre of Pakistan, 1974.
25. The building was expected to cost about four or five times as much as the other tannery structures planned in Tanzania.

CHAPTER EIGHT NOTES

1. The nine petrochemicals are: acrylonitrile, cyclohexane, isoprene, synthetic methanol, ortho-xylene, synthetic phenol, para-xylene, styrene monomer, and vinyl chloride monomer. These products, all of which are considered commodities, are used to manufacture a variety of petrochemical products, principally plastics, synthetic fibers, and synthetic rubber. Data for this chapter were obtained from questionnaires and trade journals. For a detailed description of important characteristics of the petrochemical industry, see table 1 in my "Creating a Monopoly: Product Innovation in Petrochemicals," in Richard S. Rosenbloom, ed., *Research on Technological Innovation, Management and Policy, a Research Annual, Volume 2 - 1984* (JAI Press, forthcoming).

2. The conclusion is supported by interviews, by the sizes of the coefficients in the multiple regression model (Appendix), and by the data in figure 8–3.

3. For examples of misjudgment of factors involved in foreign direct investments, see R. B. Stobaugh et al., *Nine Investments Abroad and Their Impact at Home: Case Studies in Multinational Enterprise and the U.S. Economy* (Boston: Harvard Business School, Division of Research, 1976).

4. J. M. Stopford and L. T. Wells, Jr., *Managing the Multinational Enterprise* (New York: Basic Books, 1972), p. 150.

5. R. Vernon, *Storm Over the Multinationals* (Cambridge, Mass.: Harvard University Press, 1977).

6. Seven interviews. Telesio also found this in his interviews; see Chapter 9 of this book.

7. Although I did not search explicitly for the reciprocity licensing reported by Telesio in Chapter 9 of this book, I noticed only one sale of technology for any of the nine petrochemicals that was the result of a reciprocity agreement—Du Pont's sale of cyclohexane technology to Imperial Chemical Industries in 1949.

8. R. B. Stobaugh and P. L. Townsend, "Price Forecasting and Strategic Planning: The Case of Petrochemicals," *Journal of Marketing Research,* Vol. 12 (February 1975), pp. 19–29.

9. A note on methodology: In the petrochemical industry the number of manufacturers is so highly correlated with the number of technology owners that one is forced to use just one of the numbers in an empirical study. The number of technology owners engaged in manufacturing was used as a measure of competition in the product market. This measure understates the competition in the product market to the extent that some manufacturers were using purchased technology. The number of total technology owners—manufacturing firms as well as engineering contractors—served as a measure of competition in the technology market. This measure overstates the competition in the technology market to the extent that some of the older processes were not competitive with the newer ones; but it is difficult to make judgments about the competitiveness of different processes, since it can vary from year to year and from place to place.

10. The basic argument is that under certain conditions (such as nonincreasing returns to scale), the coordination of interdependent activities by a complete set of perfectly competitive markets (which by definition contains a large number of buyers and sellers, perfect information, and so on), cannot be improved upon; thus, in such cases, there would be no advantage in replacing a perfect system of markets by a centrally administered control system. For a fuller discussion, see P. J. Buckley and M. Casson, *The Future of the Multinational Enterprise* (New York: Holmes & Meier, 1976), Chapter 2. Note that Buckley and Casson treat the loss due to a price reduction because of additional competition (my first argument) as a special case of uncertainty. For related theories: see O. E. Williamson, "The Modern Corporation: Origins, Evolution, Attributes," *Journal of Economic Literature,* Vol. XIX (December 1981), pp. 1537–1568, for a view of the corporation's vertical integra-

tion decision that emphasizes economizing on transaction costs; and A. M. Rugman, ed., *New Theories of the Multinational Enterprise* (New York: St. Martin's Press, 1982) for a discussion of the theory of internationalization that seeks to explain three modes of international operations—trade, foreign direct investment, and licensing—by multinational enterprises. For a succinct review of the literature on licensing versus investment, see R. E. Caves, *Multinational Enterprise and Economic Analysis* (Cambridge: Cambridge University Press, 1982), pp. 204–207.

The pattern in this chapter on petrochemicals is consistent with findings from a variety of U.S. industries that licensing and joint ventures become more important channels relative to wholly owned subsidiaries as the technology gets older. See E. Mansfield, A. Romeo, and S. Wagner, "Foreign Trade and U.S. Research and Development," *Review of Economics and Statistics,* LXI (February 1979), pp. 49–57; and E. Mansfield and A. Romeo, "Technology Transfer to Overseas Subsidiaries by U.S.-Based Firms," *The Quarterly Journal of Economics* (December 1980), pp. 737–750.

11. In statistical terms, the probability that small firms have a greater propensity to transfer technology through a sale rather than internal use is greater than .999 for either international or domestic transfers, with the effects of the other variables held constant; but the size of the standardized coefficient for firm size in the equation with *international* transfers of technology is twice as large as in the equation that includes *domestic* as well as international transfers. See Appendix.

12. These conclusions were not derived from one overall regression model, as were the conclusions for the licensing vs. investment decision, but were derived from several different sources: figure 8–3 in this chapter, and R. B. Stobaugh, "The Product Life Cycle, U.S. Exports, and International Investment," D.B.A. dissertation, Harvard Business School, 1968, pp. 149–50. Also, see J. M. Stopford and L. T. Wells, Jr., *Managing the Multinational Enterprise, op. cit.,* Part II.

13. These fears are described in R. Vernon, *Sovereignty at Bay* (New York: Basic Books, 1971), especially pp. 18–25, and S. Hymer, "The Efficiency (Contradictions) of Multinational Corporations," *American Economic Review* (May 1970) p. 446, and are consistent with the picture of the power of the giant firm painted by such authors as J. K. Galbraith, *The New Industrial State* (Boston: Houghton Mifflin, 1978), especially Chapter 16.

14. "World-Wide HPI Construction Boxscore," *International Hydrocarbon Processing,* February 1977.

15. For examples of firms operating in the more rapidly industrializing countries that have demonstrated a capability for producing innovations that respond to the special conditions of their own economies, see J. Fidel et al., "The Argentine Cigarette Industry: Technological Profile and Behavior," IDB/ECLA Research Programme in Science and Technology, Buenos Aires, September 1978, pp. 92, 94; C. J. Dahlman, "From Technological Dependence to Technological Development: The Case of the USIMINAS Steel Plant in Brazil," IDB/ECLA Research Programme in Science and Technology, Buenos Aires, October 1978; and J. Katz et al., "Productivity, Technology and Domestic Efforts in Research and Development," IDB/ECLA Research Programme in Science and Technology, Buenos Aires, July 1978. For evidence of the increasing capacity of some developing countries to sell plants and engineering services, see S. Lall, "Developing Countries as Exporters of Industrial Technology," *Research Policy,* Vol. 9, no. 1 (January 1980).

16. Stobaugh and Townsend, *op. cit.*

CHAPTER NINE NOTES

1. For example, Vernon estimates that U.S. multinational enterprises account for about 80 percent of total U.S. foreign direct investment. R. Vernon, *Sovereignty at Bay* (New York: Basic Books, 1971).

2. See, for example, J. Baranson, "Technology Transfer Through the International Firm," *American Economic Review,* Vol. 60 (May 1970), pp. 435–440; and D. B. Zenoff, "Licensing as a Means of Penetrating Foreign Markets," *IDEA,* Vol. 14 (Summer 1970); H. Crookell, "Licensing Technology from Multinational Firms," mimeographed. (University of Western Ontario, January 1975); and W. Wilson, "The Sale of Technology Through Licensing," a Report to the National Science Foundation (Washington, D.C., 1975); For more recent research see J. P. Killing, "Manufacturing Under License," *Business Quarterly,* (Winter 1977); D. Teece, "Technology Transfer by Multinational Firms: The Resource Cost of Transferring Technological Know-How," *The Economic Journal* (June 1977); F. Contractor, *International Technology Licensing: Compensation, Costs and Negotiations* (Lexington, Mass.: Lexington Books, 1981).

3. Multinationals based outside the United States often are comparable in size to their U.S. counterparts. In 1974, for example, 26 of the world's 50 largest firms were based outside the United States. See *Fortune World Business Directory* (New York: Time, Inc., 1975).

4. Past studies have recognized this. See Baranson, *op. cit.,* and Zenoff, *op. cit.*

5. For an exposition of the product life cycle, see R. Vernon, *op. cit.,* and L. T. Wells, Jr., *The Product Life Cycle and International Trade* (Boston: Harvard Business School, Division of Research, 1972). See also R. B. Stobaugh, "Summary of Assessment of Research Findings on U.S. International Transactions Involving Technology Transfers," Papers and Proceedings of a National Science Foundation colloquium held in Washington, D.C., November 17, 1973.

6. These benefits of licensing are also variously identified by Baranson, *op. cit.*; N. J. Behram, "Advantages and Disadvantages of Foreign Licensing," *The Patent, Trademark and Copyright Journal of Research and Education,* Vol. 2 (March 1958); L. J. Eckstrom, "Foreign Licensing—Business Considerations and Problems," reprinted from the *Proceedings of the 1959 Institute on Private Investments Abroad,* 1959; C. H. Lee, "How to Reach the Overseas Market Through Licensing," *Harvard Business Review,* Vol. 36 (January–February 1958), pp. 77–81; V. D. Travaglini, "Licensing U.S. Know-How Abroad Is Increasing," *International Commerce* (July 25, 1966), pp. 2–5; and Zenoff, *op. cit.*

7. Size of market is mentioned by Baranson, *op. cit.*; Lee, *op. cit.;* and Zenoff, *op. cit.*

8. Baranson, *op. cit.,* and Zenoff, *op. cit.*

9. See, for example, S. M. Robbins and R. B. Stobaugh, *Money in the Multinational Enterprise* (New York: Basic Books, 1973), pp. 88–90.

10. For a discussion of oligopolies based on technological innovation, see R. Vernon, *Storm Over the Multinationals* (Cambridge, Mass.: Harvard University Press, 1977), Chapter 3.

11. Baranson, *op. cit.;* Eckstrom, *op. cit.;* Lee, *op. cit.;* and Zenoff, *op. cit.*

12. Standard Industrial Classification (SIC) code for manufacturing industries, as defined by the U.S. Department of Commerce.

13. For R&D data by industry, see National Science Foundation, *Research and Development in Industry* (Washington, D.C., 1971), Tables B–3 and B–11; for industry sales (shipments) see Department of Commerce, *Annual Survey of Manufactures, 1971* (Washington, D.C., 1973), Table 1, p. 29.

14. See the Appendix to this chapter for a more detailed description of the regression model.

15. J. Schmookler, *Invention and Economic Growth* (Cambridge, Mass.: Harvard University Press, 1966), pp. 44–46, relates the number of patents pending to the R&D expenditures of U.S. companies in 18 industries; J. E. Tilton, *International Diffusion of Technology: The Case of Semiconductors* (Washington, D.C.: Brookings Institution, 1971), relates the number of patents generated in the electronics industry to expenditures on R&D.

16. In P. Telesio, *Technology Licensing and Multinational Enterprises* (New York: Praeger, 1979), on which this chapter is based, three measures of licensing are used,

each designed to capture a particular aspect of licensing. The other two measures are: 1) the sales of foreign licensees as a percentage of the total sales by controlled foreign subsidiaries plus sales by foreign licensees, and 2) the number of times individual product lines are licensed as a percentage of the total number of times individual product lines are licensed and manufactured abroad through controlled subsidiaries. Regression results are generally consistent regardless of which of these measures is used as the dependent variable; the discrepancies are discussed in the above-cited book.

17. Past studies have found a positive correlation between the number of patents generated in an industry and total R&D expenditures in that industry. See note 15.

18. See J. M. Stopford and L. T. Wells, Jr., *Managing the Multinational Enterprise* (New York: Basic Books, 1972), pp. 119–23; and Baranson, *op. cit.*

19. The Harvard Business School case "Stages of Corporate Development: A Descriptive Model" finds that companies organized into product divisions, facing separate markets, allocate resources and measure performance on the basis of return on investment, and that this ROI standard tends to be high.

20. F. T. Knickerbocker, *Oligopolistic Reaction and Multinational Enterprise* (Boston: Harvard Business School, Division of Research, 1973); and M. Dubin, "Foreign Acquisitions and the Spread of the Multinational Firm," D.B.A. dissertation, Harvard Business School, 1975. See also discussion of investment in the United States by European multinational enterprises in E. M. Graham, "Oligopolistic Imitation and European Direct Investment in the United States," D.B.A. dissertation, Harvard Business School, 1974.

21. Baranson, *op. cit.;* and D. W. Verser, "Major Policy Issues in International Petrochemical Investments: A Series of Interviews," mimeographed (Boston: Harvard Business School, April 1975). Verser observed that firms with a small international organizational structure were more willing to license abroad than were firms with large international divisions.

22. L. G. Franko, *The European Multinationals* (Stamford, Conn.: Greylock, 1976), p. 183; Graham, *op. cit.;* R. B. Stobaugh, "The Product Life Cycle, U.S. Exports, and International Investment," D.B.A. dissertation, Harvard Business School, 1968; and J. M. Stopford, "Organizing the Multinational Firm: Can the Americans Learn from the Europeans?" in M. Z. Brooke and H. L. Remmers, eds., *The Multinational Company in Europe* (Ann Arbor: University of Michigan Press, 1972).

23. Franko, *op. cit.*, pp. 207–208.

24. See R. B. Stobaugh, "Utilizing Technical Know-How in a Foreign Investment and Licensing Program," a paper delivered to the National Meeting of the Chemical Marketing Association, Houston, Texas, February 23, 1970; and L. T. Wells, Jr., "Vehicles for the International Transfer of Technology," a paper delivered to the Technology and Economic Development International Seminar, Istanbul, Turkey, May 1969.

25. Baranson, *op. cit.*, suggests licensing when the technology position of a firm is weak; Zenoff, *op. cit.* (p. 303), states that licensing may be more appropriate than investment when "the nature of a company's industrial property is such that a higher return on investment and larger profits will be derived from licensing than from other methods of market penetration."

26. See Vernon, *Storm Over the Multinationals,* Chapters 4 and 5.

27. J. W. Markham, "Concentration: A Stimulus or Retardant to Innovation?", in H. Goldschmid, M. Manne, and F. Weston, eds., *Industrial Concentration: The New Learning* (Boston: Little, Brown, 1974).

28. See Tilton, *op. cit.*

29. Stopford and Wells, *op. cit.*, Chapter 9.

30. R. B. Stobaugh et al., *Nine Investments Abroad and Their Impact at Home* (Boston: Harvard Business School, Division of Research, 1976).

31. R. B. Stobaugh et al., *Nine Investments, op. cit.*, and Robert B. Stobaugh, Piero Telesio, and José de la Torre, "U.S. Multinational Enterprises and Changes in the

Skill Composition of U.S. Employment," in Duane Kujawa, ed., *American Labor and the Multinational Corporation* (New York: Praeger, 1973).

32. Terutomo Ozawa, *Japan's Technological Challenge to the West, 1950–1974* (Cambridge, Mass.: MIT Press, 1974).

33. R. H. Mason, "Strategies of Technology Acquisition: Direct Foreign Investment vs. Unpackaged Technology," paper presented to the Southeast Asian Development Advisory Group Seminar on Multinational Corporations in Southeast Asia, Penang, Malaysia, June 24–26, 1974. This paper identifies those industries that are more likely to transfer unbundled technology to less developed countries.

CHAPTER TEN NOTES

1. For a sample of views generally favorable to the multinational enterprise, see H. G. Johnson, "The Efficiency and Welfare Implications of the International Corporation," in C. P. Kindleberger, ed., *The International Corporation: A Symposium* (Cambridge, Mass.: MIT Press, 1970); R. Vernon, *The Economic and Political Consequences of Multinational Enterprise: An Anthology* (Boston: Harvard Business School, Division of Research, 1972); R. Vernon, *Storm Over the Multinationals: The Real Issues* (Cambridge, Mass.: Harvard University Press, 1977); P. F. Drucker, "Multinationals and Developing Countries: Myths and Realities," *Foreign Affairs,* Vol. 13 (October 1974), pp. 120–134; and G. L. Reuber, *Private Foreign Investment in Development* (Oxford: Clarendon Press, 1973). For an excellent and exhaustive review of these views, see T. J. Biersteker, "Multinational Investment in Underdeveloped Countries: An Evaluation of Contending Theoretical Perspectives," Ph.D. dissertation, Massachusetts Institute of Technology, Political Science Department, 1976, Chapter 2.

2. The literature generally critical of the role of multinational enterprises has been exhaustively reviewed in Biersteker, *op. cit.,* Chapter 1. Also, see R. Barnet and R. Muller, *Global Reach: The Power of Multinational Corporations* (New York: Simon and Schuster, 1974); P. P. Streeten and S. Lall, *Foreign Investment, Transnationals, and Developing Countries* (London: Macmillan & Co., 1977); S. Hymer, "The Multinational Corporation and the Law of Uneven Development," in J. Bhagwati, *Economics and World Order: From the 1970s to the 1990s* (New York: Macmillan & Co., 1972); C. V. Vaitsos, *Intercountry Income Distribution and Transnational Enterprises* (Oxford: Clarendon Press, 1974); and C. V. Vaitsos, "Strategic Choices in the Commercialization of Technology: The Point of View of Developing Countries," *International Social Science Journal* (Summer 1973).

3. P. P. Streeten and S. Lall, *Main Findings of a Study of Private Foreign Investment in Selected Developing Countries* (TD/B/C.3/11) (Geneva, Switzerland: United Nations, Conference on Trade and Development, 1973), p. 11.

4. P. Richardson, "Business Policy in Iran," mimeographed (Tehran: Iran Center for Management Studies, July 1976).

5. J. M. Stopford and L. T. Wells, Jr., *Managing the Multinational Enterprise: Organization of the Firm and Ownership of Subsidiaries* (New York: Basic Books, 1972), pp. 113–117.

6. J. Baranson, *Manufacturing Problems in India: The Cummins Diesel Experience* (Syracuse, N.Y.: Syracuse University Press, 1967).

7. G. S. Edelberg, "The Procurement Practices of the Mexican Affiliates of Selected United States Automobile Firms," D.B.A. dissertation, Harvard Business School, 1963, p. 58.

8. D. T. Brash, *American Investment in Australian Industry* (Cambridge, Mass.: Harvard University Press, 1966), p. 206.

9. A. E. Safarian, *Foreign Ownership of Canadian Industry* (Toronto: McGraw-Hill, 1966), p. 274.

10. Reuber, *op. cit.,* pp. 221–237.
11. Biersteker, *op. cit.,* pp. 224–227.
12. Stopford and Wells, *op. cit.,* pp. 163–164.
13. Most assemblers, especially in the automotive industry, imported their required parts as a completely knock-down (CKD) kit. When a specific part was deleted from this kit and produced locally, the assembler received a credit toward the total cost of the kit. This credit, called a deletion allowance, was usually far less than the cost of the deleted part.
14. J. C. Shearer, *High-Level Manpower in Overseas Subsidiaries* (Princeton, N.J.: Princeton University, Department of Economics and Sociology, 1960), Chapter 1.
15. Shearer, *op. cit.,* Chapter 3; D. T. Brash, *op. cit.,* pp. 111–112.
16. Biersteker, *op. cit.,* p. 220.
17. Shearer, *op. cit.,* Chapter 5; Brash, *op. cit.,* p. 111.
18. Stopford and Wells, *op. cit.,* p. 166.
19. R. Vernon, *Sovereignty at Bay: The Multinational Spread of United States Enterprises* (New York: Basic Books, 1971), pp. 141–150.
20. See Chapter 2 of this book and W. A. Yeoman, "Selection of Production Process for the Manufacturing Subsidiaries of U.S.-Based Multinational Corporations," D.B.A. dissertation, Harvard University, 1968, Chapter 5.
21. Shearer, *op. cit.,* pp. 22–23, 67–71.
22. A. E. Safarian, "The Performance of Foreign-Owned Firms in Canada," Canadian-American Committee, 1969, p. 51.
23. Brash, *op. cit.,* p. 105.
24. Hymer, *op. cit.,* pp. 113–137.
25. Safarian, "Performance of Foreign-Owned Firms," *op. cit.,* p. 51; Shearer, *op. cit.,* p. 79.
26. P. P. Gabriel, *The International Transfer of Corporate Skills* (Boston: Harvard Business School, Division of Research, 1967), p. 153; Shearer, *op. cit.,* pp. 67, 108.
27. Brash, *op. cit.,* p. 105.
28. Gabriel, *op. cit.,* pp. 86–88.
29. T. Turner, "Two Refineries: A Comparative Study of Technology Transfer to the Nigerian Refining Industry," as quoted in Biersteker, *op. cit.,* p. 300.
30. Shearer, *op. cit.,* pp. 71–72.
31. Biersteker, *op. cit.,* pp. 220–223.
32. Reuber, *op. cit.,* p. 229.
33. Reserve Bank of India, "Foreign Collaboration in Indian Industry" (Bombay, 1968).
34. V. N. Balasubramanyam, *International Transfer of Technology to India* (New York: Praeger, 1973), pp. 71–76.
35. Stopford and Wells, *op. cit.,* p. 160; Vaitsos, *op. cit.,* pp. 115–116; M. Z. Brooke and H. L. Remmers, *The Strategy of Multinational Enterprise: Organization and Finance* (New York: American Elsevier Publishing Co., Inc., 1970), p. 174.
36. R. H. Mason, "The Multinational Firm and the Cost of Technology to Developing Countries," *California Management Review,* Vol. 15 (Summer 1973), pp. 9–11.
37. W. A. Chudson and L. T. Wells, Jr., *The Acquisition of Technology from Multinational Corporations by Developing Countries* (United Nations, 1974), p. 39.
38. Vaitsos, *op. cit.,* pp. 88–90.
39. *Ibid.,* Chapter 4.
40. G. Salehkhou, "Commercialization of Technology in Developing Countries: Transfer of Pharmaceutical Technology to Iran," Ph.D. dissertation, New School for Social Research, 1974, Chapter 6.
41. United Nations Conference on Trade and Development, *Major Issues Arising from the Transfer of Technology to Developing Countries* (New York: United Nations, 1975), pp. 16–17.
42. Reuber, *op. cit.,* p. 229; Brash, *op. cit.,* p. 211.

43. Brash, *op. cit.*, pp. 215-218; S. M. Robbins and R. B. Stobaugh, *Money in the Multinational Enterprise* (New York: Basic Books, 1973), pp. 91-92.
44. Stopford and Wells, *op. cit.*, pp. 159-163.
45. *Ibid.*, p. 162.
46. Brash, *op. cit.*, p. 146.
47. Reuber, *op. cit.*, pp. 229-237.
48. Streeten and Lall, *op. cit.*, p. 146.
49. Brash, *op. cit.*, Chapter 7; Safarian, "The Performance of Foreign-Owned Firms in Canada," *op. cit.*
50. Brash, *op. cit.*, pp. 173-175.
51. A. O. Krueger, *The Benefits and Costs of Import Substitution in India: A Microeconomic Study* (Minneapolis, Minn.: University of Minnesota Press, 1975), Chapter 5.
52. See Yeoman, *op. cit.*, Chapter 6; and L. T. Wells, Jr., "Economic Man and Engineering Man: Choice of Technology in a Low-Wage Country," *Public Policy*, Vol. 21 (Summer 1973), pp. 330-337.
53. See, for example, R. Vernon, *Sovereignty at Bay, op. cit.*, pp. 103-105; R. Vernon, *The Operations of United States Enterprises in Developing Countries: Their Role in Trade and Development* (New York: United Nations Conference on Trade and Development, 1972), pp. 14-17; José de la Torre, "Exports of Manufactured Goods from Developing Countries: Marketing Factors and the Role of Foreign Enterprise," D.B.A. dissertation, Harvard Business School, 1970, pp. 77-85; United Nations, *The Acquisition of Technology, op. cit.*, p. 30.
54. Stopford and Wells, *op. cit.*, pp. 159-160; Brash, *op. cit.*, pp. 95-101.

CHAPTER ELEVEN NOTES

1. See R. C. Ronstadt, *Research and Development Abroad by U.S. Multinationals* (New York: Praeger, 1977), where these data are presented at the beginnings of Chapters 2-8.
2. Six "unintentional" acquisitions of R&D units were TTUs, because the U.S. parent had supplied technology via licensing for several years to the subsidiaries where these R&D units were located. *Ibid.*, pp. 34-35.
3. *Ibid.*, Chapter 12.
4. The foregoing observations are consistent with studies showing that multinational manufacturing organizations make their first foreign investments in nations with large markets for products that have been sold first in the United States. For instance, see R. Vernon, *Sovereignty at Bay: The Multinational Spread of U.S. Enterprises* (New York: Basic Books, 1971). See especially Chapter 3, on "The Manufacturing Industries," for citations of various studies.
5. A number of other studies reinforce these observations. First, several works suggest that U.S. multinationals enjoy no advantage from venturing abroad solely to develop new products. See Y. Aharoni, *The Foreign Investment Decision Process* (Boston: Harvard Business School, Division of Research, 1966); and Vernon, *op. cit.* The opportunity to develop new products for a foreign market does not usually occur unless general management, marketing, engineering, and manufacturing resources are already established in these foreign markets, supporting themselves by producing and selling products initially developed for the U.S. market. Also, empirical and theoretical studies of the growth of the foreign subsidiaries suggest that foreign managers may seek new investments in their local markets in order to insure growth. For instance, see E. T. Penrose, "Foreign Investment and the Growth of the Firm," *Economic Journal*, Vol. 66 (June 1956), pp. 220-235.
6. T. A. Wise, "IBM's Big Gamble," *Fortune* (August 1965), pp. 67-72.
7. Vernon, *op. cit.*, Chapters 2 and 3.

8. Theoretical work dealing with the communication of technical information corroborates the need of foreign managers for close proximity to functional R&D operations. Since new products often require substantial changes in product characteristics, close and rapid communication between general managers and marketing, engineering, and manufacturing personnel is a necessary condition for successful innovation. See R. Rosenbloom and F. W. Wolek, *Technology and Information Transfer* (Boston: Harvard Business School, Division of Research, 1970), pp. 101–108; J. Morton, *Organizing for Innovation* (New York: Random House, 1969); and E. Mansfield, *Technological Change* (New York: W. W. Norton & Co., 1971), pp. 84–88.

9. Where and when Corporate Technology Units were actually established is explained by a theory of technological competition and comparative costs as these costs apply to highly skilled professional staffs. Geographic location is not determined by factors relevant to the foreign investment process in manufacturing. Rather, the explanation is based on the ability of organizations to use highly skilled labor in order to remain on the frontier of a technological field. According to the theory, an important aspect of competition is the investigation of new technological possibilities that may have important ramifications for future business. The implication is that the development of technology with potentially revolutionary consequences cannot always be performed entirely in the United States. Consequently, corporate managers may locate R&D investments abroad to reduce or avoid uncertainty about potential technological or scientific advances that may alter their businesses. These ideas are summarized in H. Johnson, "The State of Theory in Relation to Empirical Analysis," in R. Vernon, ed., *The Technology Factor in International Trade* (New York: National Bureau of Economic Research, 1970), pp. 18–19. Mansfield also discusses the research process as "uncertainty reduction" in his *Technological Change, op. cit.,* pp. 57–61.

10. Ronstadt, *op. cit.,* Chapter 8.

11. Of course, exceptions to this process of R&D investment exist, and it should not be inferred that all organizations *must* follow this sequence of evolution, or that the evolutionary process is linear and one way. Yet the evidence indicates a general tendency, at this point in time, for R&D investments to evolve in the manner described.

12. Vernon, *op. cit.* This study's findings should also interest scholars outside international areas of research. For instance, the process of R&D investment for foreign R&D activities should interest scholars involved in generating theory about the management of technological innovation. The key need is to determine if a particular R&D investment process exists for domestic R&D investments, since they still constitute the lion's share of R&D spending. How could such a process affect existing findings about the management, staffing, and organization of R&D labs? For instance, are technological "gatekeepers" more or less critical in particular kinds of R&D investments? Are potential technological entrepreneurs better assigned to a particular kind of R&D investment early in their careers? And so on. Similarly, scholars interested in theories of economic development and the role of technological innovation may find the notions of specific kinds of R&D investments, and of the R&D investment process, useful ones for research purposes.

CHAPTER TWELVE NOTES

1. H. J. Leavitt, *Managerial Psychology,* 3rd ed. (Chicago: University of Chicago Press, 1972).

2. For systems theory, see L. Von Bertalanffy, "General Systems: A Critical Review," in *Yearbook of the Society for General Systems Research,* Vol. VII (1962), pp. 1–20; J. C. Emery, *Organizational Planning and Control Systems* (London: Macmillan,

1971); and A. Rapoport, "Mathematical Aspects of General Systems Analysis," in *The Sciences of Man: Problems and Orientations* (The Hague: UNESCO, 1968). For sociology and organizational behavior, see D. S. Pugh, D. J. Hickson, C. R. Hinings, K. M. MacDonald, and C. Turner, "A Conceptual Scheme for Organizational Analysis," *Administrative Science Quarterly,* Vol. 8 (December 1964), pp. 289–315; P. Lawrence and J. Lorsch, *Organization and Environment* (Boston: Harvard Business School, Division of Research, 1967); and L. B. Mohr, "Organizational Technology and Organizational Structure," *Administrative Science Quarterly,* Vol. 16 (December 1971), pp. 444–459. For management control, see R. N. Anthony, *Planning and Control Systems: A Framework for Analysis* (Boston: Harvard Business School, Division of Research, 1965); and R. N. Anthony, J. Dearden, and R. F. Vancil, *Management Control Systems* (Homewood, Ill.: Richard D. Irwin, 1972). For international business, see Y. Aharoni, *The Foreign Investment Decision Process* (Boston: Harvard Business School, Division of Research, 1966); L. G. Franko, *The European Multinationals: A Renewed Challenge to American and British Big Business* (Stamford, Conn.: Greylock, 1976); S. Robbins and R. B. Stobaugh, *Money in the Multinational Enterprise* (New York: Basic Books, 1974); J. M. Stopford and L. T. Wells, Jr., *Managing the Multinational Enterprise* (New York: Basic Books, 1974); and R. Vernon, *Sovereignty at Bay* (New York: Basic Books, 1972). For business policy, see J. L. Bower, *Managing the Resource Allocation Process* (Boston: Harvard Business School, Division of Research, 1970).

3. E. R. Barlow, *Management of Foreign Manufacturing Subsidiaries* (Boston: Harvard Business School, Division of Research, 1953); "Worldwide Manufacturing Policy," *Business International,* 1968; W. Skinner, *The Management of International Manufacturing* (New York: John Wiley and Sons, 1968).

4. See C. W. Skinner and D. C. D. Rogers, *Manufacturing Policy in the Plastics Industry: A Casebook of Major Production Problems,* 3rd ed. (Homewood, Ill.: Richard D. Irwin, 1968); and C. W. Skinner, "Manufacturing—Missing Link in Corporate Strategy," *Harvard Business Review,* Vol. 47, (May–June 1969), pp. 136–145.

5. In some firms, interviews were held at more than one location. Thus, in total, executives were interviewed at forty-two locations, of which twenty-two were corporate headquarters (three in the United States and nineteen in France). The remaining interviews, all held in France, were at four area headquarters (all U.S. firms), three product division headquarters (one U.S. firm and two French), and thirteen subsidiaries (eleven U.S. firms, one French, and one U.K./Netherlands).

6. R. B. Stobaugh and P. L. Townsend, "Price Forecasting and Strategic Planning: The Case of Petrochemicals," *Journal of Marketing Research,* Vol. 12 (February 1975), pp. 19–29. The impact of a subsidiary's receiving the benefit of varying portions of the parent's accumulated production is tested by a simple computer model in R. B. Stobaugh, *U.S. Taxation of United States Manufacturing Abroad: Likely Effects of Taxing Unremitted Profits* (New York: Financial Executives Research Foundation, 1976), p. 66.

7. Unless otherwise noted, this and subsequent descriptions of individual firms are based on the author's field interviews.

8. M. Y. Yoshino, *Japan's Managerial System* (Cambridge, Mass.: MIT Press, 1971), pp. 254–272.

9. H. J. Leavitt, *Managerial Psychology, op. cit.*

10. J. Woodward, *Industrial Organization: Theory and Practice* (London: Oxford University Press, 1965), Chapter 3.

11. R. J. Alsegg, *Control Relationships Between American Corporations and Their European Subsidiaries* (New York: American Management Association, 1971).

Index

317